Statistics Demystified

Demystified Series

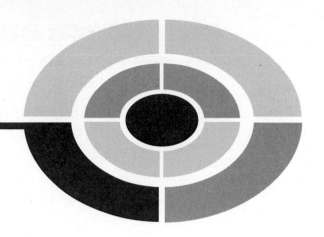

Statistics Demystified

STAN GIBILISCO

McGRAW-HILL
New York   Chicago   San Francisco   Lisbon   London
Madrid   Mexico City   Milan   New Delhi   San Juan
Seoul   Singapore   Sydney   Toronto

The McGraw·Hill Companies

Cataloging-in-Publication Data is on file with the Library of Congress

4 5 6 7 8 9 0   DOC/DOC   0 1 0 9 8 7 6 5

ISBN 0-07-143118-7

The sponsoring editor for this book was Judy Bass and the production supervisor was Pamela A. Pelton. It was set in Times Roman by Keyword Publishing Services, Ltd. The art director for the cover was Margaret Webster-Shapiro; the cover designer was Handel Low.

Printed and bound by RR Donnelley.

McGraw-Hill books are available at special quantity discounts to use as premiums and sales promotions, or for use in corporate training programs. For more information, please write to the Director of Special Sales, McGraw-Hill Professional, Two Penn Plaza, New York, NY 10121-2298. Or contact your local bookstore.

 This book is printed on recycled, acid-free paper containing a minimum of 50% recycled, de-inked fiber.

To Tim, Tony, and Samuel from Uncle Stan

CONTENTS

CONTENTS

PREFACE

This book is for people who want to learn or review the basic concepts and definitions in statistics and probability at the high-school level. It can serve as a supplemental text in a classroom, tutored, or home-schooling environment. It should be useful for career changers who need to refresh their knowledge. I recommend that you start at the beginning of this book and go straight through.

Many students have a hard time with statistics. This is a preparatory text that can get you ready for a standard course in statistics. If you've had trouble with other statistics books because they're too difficult, or because you find yourself performing calculations without any idea of what you're really doing, this book should help you.

This book contains numerous practice quiz, test, and exam questions. They resemble the sorts of questions found in standardized tests. There is a short quiz at the end of every chapter. The quizzes are "open-book." You may (and should) refer to the chapter texts when taking them. When you think you're ready, take the quiz, write down your answers, and then give your list of answers to a friend. Have the friend tell you your score, but not which questions you got wrong. The answers are listed in the back of the book. Stick with a chapter until you get most of the answers correct.

This book is divided into two multi-chapter parts, followed by part tests. Take these tests when you're done with the respective parts and have taken all the chapter quizzes. The part tests are "closed-book," but the questions are easier than those in the quizzes. A satisfactory score is 75% or more correct. Again, answers are in the back of the book.

There is a final exam at the end of this course. It contains questions drawn uniformly from all the chapters in the book. Take it when you have finished both parts, both part tests, and all of the chapter quizzes. A satisfactory score is at least 75% correct answers. With the part tests and the final exam, as with the quizzes, have a friend tell you your score without letting you know which questions you missed. That way, you will not subconsciously memorize the answers. You can check to see where your knowledge is strong and where it is not.

I recommend that you complete one chapter every couple of weeks. An hour or two daily ought to be enough time for this. When you're done with the course, you can use this book, with its comprehensive index, as a permanent reference.

Suggestions for future editions are welcome.

STAN GIBILISCO

ACKNOWLEDGMENTS

Illustrations in this book were generated with *CorelDRAW*. Some of the clip art is courtesy of Corel Corporation.

I extend thanks to Steve Sloat and Tony Boutelle, who helped with the technical evaluation of the manuscript.

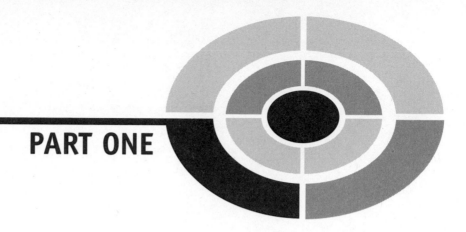

PART ONE

Statistical Concepts

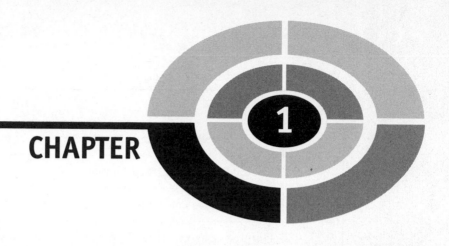

CHAPTER 1

Background Math

This chapter is a review of basic mathematical principles. Some of this is abstract if considered in theoretical isolation, but when it comes to knowing what we're talking about in statistics and probability, it's important to be familiar with sets, number theory, relations, functions, equations, and graphs.

Table 1-1 lists some of the symbols used in mathematics. Many of these symbols are also encountered in statistics. It's a good idea to become familiar with them.

Sets

A *set* is a collection or group of definable *elements* or *members*. Some examples of set elements are:

- points on a line
- instants in time

Table 1-1   Symbols used in basic mathematics.

Symbol	Description
{ }	Braces; objects between them are elements of a set
$\Rightarrow$	Logical implication; read "implies"
$\Leftrightarrow$	Logical equivalence; read "if and only if"
$\forall$	Universal quantifier; read "for all" or "for every"
$\exists$	Existential quantifier; read "for some"
$\mid$ ∶	Logical expression; read "such that"
&	Logical conjunction; read "and"
N	The set of natural numbers
Z	The set of integers
Q	The set of rational numbers
R	The set of real numbers
$\aleph$	Transfinite (or infinite) cardinal number
$\varnothing$	The set with no elements; read "the empty set" or "the null set"
$\cap$	Set intersection; read "intersect"
$\cup$	Set union; read "union"
$\subset$	Proper subset; read "is a proper subset of"
$\subseteq$	Subset; read "is a subset of"
$\in$	Element; read "is an element of" or "is a member of"
$\notin$	Non-element; read "is not an element of" or "is not a member of"

$=$	Equality; read "equals" or "is equal to"
$\neq$	Not-equality; read "does not equal" or "is not equal to"
$\approx$	Approximate equality; read "is approximately equal to"
$<$	Inequality; read "is less than"
$\leq$	Equality or inequality; read "is less than or equal to"
$>$	Inequality; read "is greater than"
$\geq$	Equality or inequality; read "is greater than or equal to"
$+$	Addition; read "plus"
$-$	Subtraction, read "minus"
$\times$. $*$	Multiplication; read "times" or "multiplied by"
$\div$ /	Quotient; read "over" or "divided by"
:	Ratio or proportion; read "is to"
!	Product of all natural numbers from 1 up to a certain value; read "factorial"
( )	Quantification; read "the quantity"
[ ]	Quantification; used outside ( )
{ }	Quantification; used outside [ ]

- individual apples in a basket
- coordinates in a plane
- coordinates in space
- coordinates on a display
- curves on a graph or display
- chemical elements
- individual people in a city
- locations in memory or storage
- data bits, bytes, or characters
- subscribers to a network

If an object or number (call it a) is an element of set A, this fact is written as:

$$a \in A$$

The $\in$ symbol means "is an element of."

SET INTERSECTION

The *intersection* of two sets A and B, written $A \cap B$, is the set C such that the following statement is true for every element x:

$$x \in C \text{ if and only if } x \in A \text{ and } x \in B$$

The $\cap$ symbol is read "intersect."

SET UNION

The *union* of two sets A and B, written $A \cup B$, is the set C such that the following statement is true for every element x:

$$x \in C \text{ if and only if } x \in A \text{ or } x \in B$$

The $\cup$ symbol is read "union."

SUBSETS

A set A is a *subset* of a set B, written $A \subseteq B$, if and only if the following holds true:

$$x \in A \text{ implies that } x \in B$$

The $\subseteq$ symbol is read "is a subset of." In this context, "implies that" is meant in the strongest possible sense. The statement "This implies that" is equivalent to "If this is true, then that is always true."

PROPER SUBSETS

A set A is a *proper subset* of a set B, written $A \subset B$, if and only if the following both hold true:

$$x \in A \text{ implies that } x \in B$$
$$\text{as long as } A \neq B$$

The $\subset$ symbol is read "is a proper subset of."

DISJOINT SETS

Two sets A and B are *disjoint* if and only if all three of the following conditions are met:

$$A \neq \varnothing$$
$$B \neq \varnothing$$
$$A \cap B = \varnothing$$

where $\varnothing$ denotes the *empty set*, also called the *null set*. It is a set that doesn't contain any elements, like a basket of apples without the apples.

COINCIDENT SETS

Two non-empty sets A and B are *coincident* if and only if, for all elements x, both of the following are true:

$$x \in A \text{ implies that } x \in B$$
$$x \in B \text{ implies that } x \in A$$

Relations and Functions

Consider the following statements. Each of them represents a situation that could occur in everyday life.

- The outdoor air temperature varies with the time of day.
- The time the sun is above the horizon on June 21 varies with the latitude of the observer.
- The time required for a wet rag to dry depends on the air temperature.

All of these expressions involve something that depends on something else. In the first case, a statement is made concerning temperature versus time; in the second case, a statement is made concerning sun-up time versus latitude; in the third case, a statement is made concerning time versus temperature. Here, the term *versus* means "compared with."

INDEPENDENT VARIABLES

An *independent variable* changes, but its value is not influenced by anything else in a given scenario. Time is often treated as an independent variable. A lot of things depend on time.

When two or more variables are interrelated, at least one of the variables is independent, but they are not all independent. A common and simple situation is one in which there are two variables, one of which is independent. In the three situations described above, the independent variables are time, latitude, and air temperature.

DEPENDENT VARIABLES

A *dependent variable* changes, but its value is affected by at least one other factor in a situation. In the scenarios described above, the air temperature, the sun-up time, and time are dependent variables.

When two or more variables are interrelated, at least one of them is dependent, but they cannot all be dependent. Something that's an independent variable in one instance can be a dependent variable in another case. For example, the air temperature is a dependent variable in the first situation described above, but it is an independent variable in the third situation.

SCENARIOS ILLUSTRATED

The three scenarios described above lend themselves to illustration. In order, they are shown crudely in Fig. 1-1.

Figure 1-1A shows an example of outdoor air temperature versus time of day. Drawing B shows the sun-up time (the number of hours per day in which the sun is above the horizon) versus latitude on June 21, where points

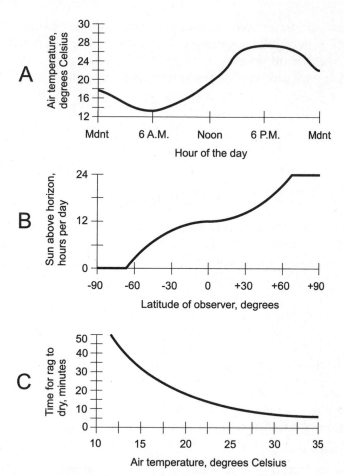

Fig. 1-1.   Three "this-versus-that" scenarios. At A, air temperature versus time of day; at B, sun-up time versus latitude; at C, time for rag to dry versus air temperature.

south of the equator have negative latitude and points north of the equator have positive latitude. Drawing C shows the time it takes for a rag to dry, plotted against the air temperature.

The scenarios represented by Figs. 1-1A and C are fiction, having been contrived for this discussion. But Fig. 1-1B represents a physical reality; it is true astronomical data for June 21 of every year on earth.

WHAT IS A RELATION?

All three of the graphs in Fig. 1-1 represent *relations*. In mathematics, a relation is an expression of the way two or more variables compare or interact. (It could just as well be called a relationship, a comparison, or an interaction.) Figure 1-1B, for example, is a graph of the relation between the latitude and the sun-up time on June 21.

When dealing with relations, the statements are equally valid if the variables are stated the other way around. Thus, Fig. 1-1B shows a relation between the sun-up time on June 21 and the latitude. In a relation, "this versus that" means the same thing as "that versus this." Relations can always be expressed in graphical form.

WHEN IS A RELATION A FUNCTION?

A *function* is a special type of mathematical relation. A relation describes how variables compare with each other. In a sense, it is "passive." A function transforms, processes, or morphs the quantity represented by the independent variable into the quantity represented by the dependent variable. A function is "active."

All three of the graphs in Fig. 1-1 represent functions. The changes in the value of the independent variable can, in some sense, be thought of as causative factors in the variations of the value of the dependent variable. We might re-state the scenarios this way to emphasize that they are functions:

- The outdoor air temperature is a function of the time of day.
- The sun-up time on June 21 is a function of the latitude of the observer.
- The time required for a wet rag to dry is a function of the air temperature.

A relation can be a function only when every element in the set of its independent variables has at most one correspondent in the set of dependent variables. If a given value of the dependent variable in a relation has more than one independent-variable value corresponding to it, then that relation might nevertheless be a function. But if any given value of the independent variable corresponds to more than one dependent-variable value, that relation is not a function.

REVERSING THE VARIABLES

In graphs of functions, independent variables are usually represented by horizontal axes, and dependent variables are usually represented by vertical axes. Imagine a movable, vertical line in a graph, and suppose that you can move it back and forth. A curve represents a function if and only if it never intersects the movable vertical line at more than one point.

Imagine that the independent and dependent variables of the functions shown in Fig. 1-1 are reversed. This results in some weird assertions:

- The time of day is a function of the outdoor air temperature.
- The latitude of an observer is a function of the sun-up time on June 21.
- The air temperature is a function of the time it takes for a wet rag to dry.

The first two of these statements are clearly ridiculous. Time does not depend on temperature. You can't make time go backwards by cooling things off or make it rush into the future by heating things up. Your geographic location is not dependent on how long the sun is up. If that were true, you would be at a different latitude a week from now than you are today, even if you don't go anywhere (unless you live on the equator!).

If you turn the graphs of Figs. 1-1A and B sideways to reflect the transposition of the variables and then perform the vertical-line test, you'll see that they no longer depict functions. So the first two of the above assertions are not only absurd, they are false.

Figure 1-1C represents a function, at least in theory, when "stood on its ear." The statement is still strange, but it can at least be true under certain conditions. The drying time of a standard-size wet rag made of a standard material could be used to infer air temperature experimentally (although humidity and wind speed would be factors too). When you want to determine whether or not a certain graph represents a mathematical function, use the vertical-line test, not the common-sense test!

DOMAIN AND RANGE

Let f be a function from set A to set B. Let A' be the set of all elements a in A for which there is a corresponding element b in B. Then A' is called the *domain* of f.

Let f be a function from set A to set B. Let B' be the set of all elements b in B for which there is a corresponding element a in A. Then B' is called the *range* of f.

PROBLEM 1-1

Figure 1-2 is called a *Venn diagram*. It shows two sets A and B, and three points or elements P, Q, and R. What is represented by the cross-hatched region? Which of the points, if any, is in the intersection of sets A and B? Which points, if any, are in the union of sets A and B?

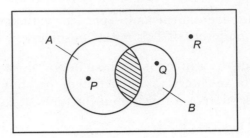

Fig. 1-2.   Illustration for Problem 1-1.

SOLUTION 1-1

The cross-hatched region represents all the elements that are in both set A and set B, so it is an illustration of $A \cap B$, the intersection of A and B. None of the elements shown are in $A \cap B$. Points P and Q are in $A \cup B$, the union of A and B.

PROBLEM 1-2

Figure 1-3 is an illustration of a relation that maps certain points in a set C to certain points in a set D. Only those points shown are involved in this relation. Is this relation a function? If so, how can you tell? If not, why not?

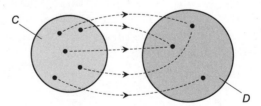

Fig. 1-3.   Illustration for Problem 1-2.

SOLUTION 1-2

The relation is a function, because each value of the independent variable, shown by the points in set C, maps into at most one value of the dependent variable, represented by the points in set D.

Numbers

A *number* is an abstract expression of a quantity. Mathematicians define numbers in terms of sets containing sets. All the known numbers can be built up from a starting point of zero. *Numerals* are the written symbols that are agreed-on to represent numbers.

NATURAL AND WHOLE NUMBERS

The *natural numbers*, also called *whole numbers* or *counting numbers*, are built up from a starting point of 0 or 1, depending on which text you consult. The set of natural numbers is denoted N. If we include 0, we have this:

$$N = \{0, 1, 2, 3, \ldots, n, \ldots\}$$

In some instances, 0 is not included, so:

$$N = \{1, 2, 3, 4, \ldots, n, \ldots\}$$

Natural numbers can be expressed as points along a geometric ray or half-line, where quantity is directly proportional to displacement (Fig. 1-4).

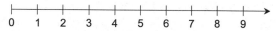

Fig. 1-4.   The natural numbers can be depicted as points on a half-line or ray.

INTEGERS

The set of natural numbers, including zero, can be duplicated and inverted to form an identical, mirror-image set:

$$-N = \{0, -1, -2, -3, \ldots, -n, \ldots\}$$

The union of this set with the set of natural numbers produces the set of *integers*, commonly denoted Z:

$$Z = N \cup -N$$
$$= \{\ldots, -n, \ldots, -2, -1, 0, 1, 2, \ldots, n, \ldots\}$$

Integers can be expressed as individual points spaced at equal intervals along a line, where quantity is directly proportional to displacement (Fig. 1-5). In the illustration, integers correspond to points where hash marks cross the line. The set of natural numbers is a proper subset of the set of integers:

$$N \subset Z$$

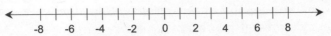

Fig. 1-5. The integers can be depicted as individual points, spaced at equal intervals on a line. The real numbers can be depicted as the set of all points on the line.

For any number a, if a is an element of N, then a is an element of Z. The converse of this is not true. There are elements of Z (namely, the negative integers) that are not elements of N.

RATIONAL NUMBERS

A *rational number* (the term derives from the word *ratio*) is a quotient of two integers, where the denominator is positive. The standard form for a rational number r is:

$$r = a/b$$

Any such quotient is a rational number. The set of all possible such quotients encompasses the entire set of rational numbers, denoted Q. Thus:

$$Q = \{x \mid x = a/b\}$$

where $a \in Z$, $b \in Z$, and $b > 0$. (Here, the vertical line means "such that.") The set of integers is a proper subset of the set of rational numbers. Thus, the natural numbers, the integers, and the rational numbers have the following relationship:

$$N \subset Z \subset Q$$

DECIMAL EXPANSIONS

Rational numbers can be denoted in decimal form as an integer followed by a period (radix point) followed by a sequence of digits. The digits following the radix point always exist in either of two forms:

- a finite string of digits beyond which all digits are zero
- an infinite string of digits that repeat in cycles

Examples of the first type, known as *terminating decimals*, are:

$$3/4 = 0.750000\ldots$$

$$-9/8 = -1.1250000\ldots$$

Examples of the second type, known as *nonterminating, repeating decimals*, are:

$$1/3 = 0.33333\ldots$$

$$-123/999 = -0.123123123\ldots$$

IRRATIONAL NUMBERS

An *irrational number* is a number that cannot be expressed as the ratio of any two integers. (This is where the term "irrational" comes from; it means "existing as no ratio.") Examples of irrational numbers include:

- the length of the diagonal of a square in a flat plane that is 1 unit long on each edge; this is $2^{1/2}$, also known as the square root of 2
- the circumference-to-diameter ratio of a circle as determined in a flat plane, conventionally named by the lowercase Greek letter pi (π)

All irrational numbers share one quirk: they cannot be expressed precisely using a radix point. When an attempt is made to express such a number in this form, the result is a *nonterminating, nonrepeating* decimal. No matter how many digits are specified to the right of the radix point, the expression is only an approximation of the actual value of the number. The best we can do is say things like this:

$$2^{1/2} = 1.41421356\ldots$$

$$\pi = 3.14159\ldots$$

Sometimes, "squiggly equals signs" are used to indicate that values are approximate:

$$2^{1/2} \approx 1.41421356$$

$$\pi \approx 3.14159$$

The set of irrational numbers can be denoted S. This set is entirely disjoint from the set of rational numbers, even though, in a sense, the two sets are intertwined:

$$S \cap Q = \varnothing$$

REAL NUMBERS

The set of real numbers, denoted **R**, is the union of the sets of rational and irrational numbers:

$$R = Q \cup S$$

For practical purposes, **R** can be depicted as a continuous geometric line, as shown in Fig. 1-5. (In theoretical mathematics, the assertion that all the points on a geometric line correspond one-to-one with the real numbers is known as the *Continuum Hypothesis*. It may seem obvious to the lay person that this ought to be true, but proving it is far from trivial.)

The set of real numbers is related to the sets of rational numbers, integers, and natural numbers as follows:

$$N \subset Z \subset Q \subset R$$

The operations of addition, subtraction, multiplication, and division can be defined over the set of real numbers. If # represents any one of these operations and x and y are elements of **R** with $y \neq 0$, then:

$$x \# y \in R$$

PROBLEM 1-3
Given any two different rational numbers, is it always possible to find another rational number between them? That is, if x and y are any two different rational numbers, is there always some rational number z such that $x < z < y$ (x is less than z and z is less than y)?

SOLUTION 1-3
Yes. We can prove this by arming ourselves with the general formula for the sum of two fractions:

$$a/b + c/d = (ad + bc)/(bd)$$

where neither b nor d is equal to 0. Suppose we have two rational numbers x and y, consisting of ratios of integers a, b, c, and d, such that:

$$x = a/b$$

$$y = c/d$$

We can find the *arithmetic mean*, also called the *average*, of these two rational numbers; this is the number z we seek. The arithmetic mean of two numbers is equal to half the sum of the two numbers.

The arithmetic mean of any two rational numbers is always another rational number. This can be proven by noting that:

$$(x + y)/2 = (a/b + c/d)/2$$
$$= (ad + bc)/(2bd)$$

The product of any two integers is another integer. Also, the sum of any two integers is another integer. Thus, because a, b, c, and d are integers, we know that $ad + bc$ is an integer, and also that $2bd$ is an integer. Call these derived integers p and q, as follows:

$$p = ad + bc$$
$$q = 2bd$$

The arithmetic mean of x and y is equal to p/q, which is a rational number because it is equal to the ratio of two integers.

One-Variable Equations

The objective of solving a single-variable equation is to get it into a form where the expression on the left-hand side of the equals sign is the variable being sought (for example, x) standing all alone, and the expression on the right-hand side of the equals sign is an expression that does not contain the variable being sought.

ELEMENTARY RULES

There are several ways in which an equation in one variable can be manipulated to obtain a solution, assuming a solution exists. The following rules can be applied in any order, and any number of times.

Addition of a quantity to each side: Any defined constant, variable, or expression can be added to both sides of an equation, and the result is equivalent to the original equation.

Subtraction of a quantity from each side: Any defined constant, variable, or expression can be subtracted from both sides of an equation, and the result is equivalent to the original equation.

Multiplication of each side by a quantity: Both sides of an equation can be multiplied by a defined constant, variable, or expression, and the result is equivalent to the original equation.

Division of each side by a quantity: Both sides of an equation can be divided by a nonzero constant, by a variable that cannot attain a value of zero, or by an expression that cannot attain a value of zero over the range of its variable(s), and the result is equivalent to the original equation.

BASIC EQUATION IN ONE VARIABLE

Consider an equation of the following form:

$$ax + b = cx + d$$

where a, b, c, and d are real-number constants, x is a variable, and $a \neq c$. This equation is solved for x as follows:

$$ax + b = cx + d$$
$$ax = cx + d - b$$
$$ax - cx = d - b$$
$$(a - c)x = d - b$$
$$x = (d - b)/(a - c)$$

FACTORED EQUATIONS IN ONE VARIABLE

Consider an equation of the following form:

$$(x - a_1)(x - a_2)(x - a_3) \cdots (x - a_n) = 0$$

where a_1, a_2, a_3, ..., a_n are real-number constants, and x is a variable. There are multiple solutions to this equation. Call the solutions x_1, x_2, x_3, and so on up to x_n, as follows:

$$x_1 = a_1$$
$$x_2 = a_2$$
$$x_3 = a_3$$
$$\downarrow$$
$$x_n = a_n$$

The *solution set* of this equation is $\{a_1, a_2, a_3, \ldots, a_n\}$.

QUADRATIC EQUATIONS

Consider an equation of the following form:

$$ax^2 + bx + c = 0$$

where a, b, and c are real-number constants, x is a variable, and a is not equal to 0. This is called the *standard form* of a *quadratic equation*. It may have no real-number solutions for x, or a single real-number solution, or two real-number solutions. The solutions of this equation, call them x_1 and x_2, can be found according to the following formulas:

$$x_1 = [-b + (b^2 - 4ac)^{1/2}]/2a$$
$$x_2 = [-b - (b^2 - 4ac)^{1/2}]/2a$$

Sometimes these are written together as a single formula, using a plus-or-minus sign ($\pm$) to indicate that either addition or subtraction can be performed. This is the well-known *quadratic formula* from elementary algebra:

$$x = [-b \pm (b^2 - 4ac)^{1/2}]/2a$$

PROBLEM 1-4
Find the solution of the following equation:

$$3x - 5 = 2x$$

SOLUTION 1-4
This equation can be put into the form $ax + b = cx + d$, where $a = 3$, $b = -5$, $c = 2$, and $d = 0$. Then, according to the general solution derived above:

$$x = (d - b)/(a - c)$$
$$= [0 - (-5)]/(3 - 2)$$
$$= 5/1$$
$$= 5$$

PROBLEM 1-5
Find the real-number solution or solutions of the following equation:

$$x^2 - x = 2$$

SOLUTION 1-5
This is a quadratic equation. It can be put into the standard quadratic form by subtracting 2 from each side:

$$x^2 - x - 2 = 0$$

There are two ways to solve this. First, note that it can be factored into the following form:

$$(x + 1)(x - 2) = 0$$

From this, it is apparent that there are two solutions for x: $x_1 = -1$ and $x_2 = 2$. Either of these will satisfy the equation because they render the first and second terms, respectively, equal to zero.

The other method of solving this equation is to use the quadratic formula, once the equation has been reduced to standard form. In this case, the constants are $a = 1$, $b = -1$, and $c = -2$. Thus:

$$x = [-b \pm (b^2 - 4ac)^{1/2}]/2a$$
$$= \{1 \pm [1^2 - 4 \times 1 \times (-2)]^{1/2}\}/(2 \times 1)$$
$$= \{1 \pm [1 - (-8)]^{1/2}\}/2$$
$$= (1 \pm 9^{1/2})/2$$
$$= (1 \pm 3)/2$$

This gives two solutions for x: $x_1 = (1 + 3)/2 = 4/2 = 2$, and $x_2 = (1 - 3)/2 = -1$. These are the same two solutions as are obtained by factoring. (It doesn't matter that they turn up in the opposite order in these two solution processes.)

Simple Graphs

When the variables in a function are clearly defined, or when they can attain only specific values (called *discrete values*), graphs can be rendered simply. Here are some of the most common types.

SMOOTH CURVES

Figure 1-6 is a graph showing two curves, each of which represents the fluctuations in the prices of a hypothetical stock during the better part of a business day. Let's call the stocks Stock X and Stock Y. Both of the curves represent functions of time. You can determine this using the vertical-line test. Neither of the curves intersects a movable, vertical line more than once.

Suppose, in the situation shown by Fig. 1-6, the stock price is considered the independent variable, and time is considered the dependent variable. To illustrate this, plot the graphs by "standing the curves on their ears," as

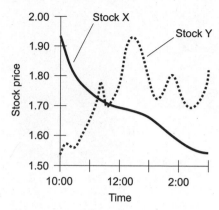

Fig. 1-6. The curves show fluctuations in the prices of hypothetical stocks during the course of a business day.

shown in Fig. 1-7. (The curves are rotated 90 degrees counterclockwise, and then mirrored horizontally.) Using the vertical-line test, it is apparent that time can be considered a function of the price of Stock X, but not a function of the price of Stock Y.

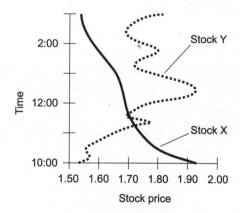

Fig. 1-7. A smooth-curve graph in which stock price is the independent variable, and time is the dependent variable.

VERTICAL BAR GRAPHS

In a *vertical bar graph*, the independent variable is shown on the horizontal axis and the dependent variable is shown on the vertical axis. Function values are portrayed as the heights of bars having equal widths. Figure 1-8 is a

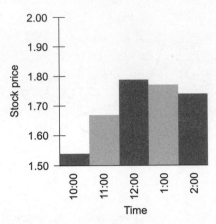

Fig. 1-8. Vertical bar graph of hypothetical stock price versus time.

vertical bar graph of the price of the hypothetical Stock Y at intervals of 1 hour.

HORIZONTAL BAR GRAPHS

In a *horizontal bar graph*, the independent variable is shown on the vertical axis and the dependent variable is shown on the horizontal axis. Function values are portrayed as the widths of bars having equal heights. Figure 1-9 is a horizontal bar graph of the price of the hypothetical Stock Y at intervals of 1 hour.

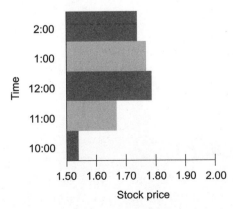

Fig. 1-9. Horizontal bar graph of hypothetical stock price versus time.

HISTOGRAMS

A *histogram* is a bar graph applied to a special situation called a *distribution*. An example is a portrayal of the grades a class receives on a test, such as is shown in Fig. 1-10. Here, each vertical bar represents a letter grade (A, B, C, D, or F). The height of the bar represents the percentage of students in the class receiving that grade.

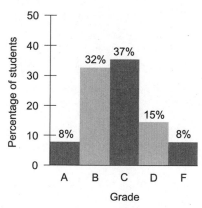

Fig. 1-10.   A histogram is a specialized form of bar graph.

In Fig. 1-10, the values of the dependent variable are written at the top of each bar. In this case, the percentages add up to 100%, based on the assumption that all of the people in the class are present, take the test, and turn in their papers. The values of the dependent variable are annotated this way in some bar graphs. It's a good idea to write in these numbers if there aren't too many bars in the graph, but it can make the graph look messy or confusing if there are a lot of bars.

Some histograms are more flexible than this, allowing for variable bar widths as well as variable bar heights. We'll see some examples of this in Chapter 4. Also, in some bar graphs showing percentages, the values do not add up to 100%. We'll see an example of this sort of situation a little later in this chapter.

POINT-TO-POINT GRAPHS

In a *point-to-point graph*, the scales are similar to those used in continuous-curve graphs such as Figs. 1-6 and 1-7. But the values of the function in a point-to-point graph are shown only for a few selected points, which are connected by straight lines.

In the point-to-point graph of Fig. 1-11, the price of Stock Y (from Fig. 1-6) is plotted on the half-hour from 10:00 A.M. to 3:00 P.M. The resulting "curve" does not exactly show the stock prices at the in-between times. But overall, the graph is a fair representation of the fluctuation of the stock over time.

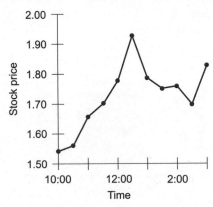

Fig. 1-11. A point-to-point graph of hypothetical stock price versus time.

When plotting a point-to-point graph, a certain minimum number of points must be plotted, and they must all be sufficiently close together. If a point-to-point graph showed the price of Stock Y at hourly intervals, it would not come as close as Fig. 1-11 to representing the actual moment-to-moment stock-price function. If a point-to-point graph showed the price at 15-minute intervals, it would come closer than Fig. 1-11 to the moment-to-moment stock-price function.

CHOOSING SCALES

When composing a graph, it's important to choose sensible scales for the dependent and independent variables. If either scale spans a range of values much greater than necessary, the *resolution* (detail) of the graph will be poor. If either scale does not have a large enough span, there won't be enough room to show the entire function; some of the values will be "cut off."

PROBLEM 1-6
Figure 1-12 is a hypothetical bar graph showing the percentage of the work force in a certain city that calls in sick on each day during a particular work week. What, if anything, is wrong with this graph?

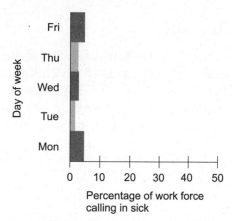

Fig. 1-12.   Illustration for Problems 1-6 and 1-7.

SOLUTION 1-6
The horizontal scale is much too large. It makes the values in the graph difficult to ascertain. It would be better if the horizontal scale showed values only in the range of 0 to 10%. The graph could also be improved by listing percentage numbers at the right-hand side of each bar.

PROBLEM 1-7
What's going on with the percentage values depicted in Fig. 1-12? It is apparent that the values don't add up to 100%. Shouldn't they?

SOLUTION 1-7
No. If they did, it would be a coincidence (and a bad reflection on the attitude of the work force in that city during that week). This is a situation in which the sum of the percentages in a bar graph does not have to be 100%. If everybody showed up for work every day for the whole week, the sum of the percentages would be 0, and Fig. 1-12 would be perfectly legitimate showing no bars at all.

Tweaks, Trends, and Correlation

Graphs can be approximated or modified by "tweaking." Certain characteristics can also be noted, such as trends and correlation. Here are a few examples.

LINEAR INTERPOLATION

The term *interpolate* means "to put between." When a graph is incomplete, estimated data can be put in the gap(s) in order to make the graph look complete. An example is shown in Fig. 1-13. This is a graph of the price of the hypothetical Stock Y from Fig. 1-6, but there's a gap during the noon hour. We don't know exactly what happened to the stock price during that hour, but we can fill in the graph using *linear interpolation*. A straight line is placed between the end points of the gap, and then the graph looks complete.

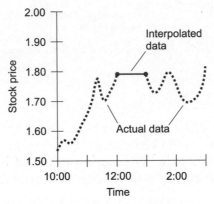

Fig. 1-13. An example of linear interpolation. The thin solid line represents the interpolation of the values for the gap in the actual available data (heavy dashed curve).

Linear interpolation almost always produces a somewhat inaccurate result. But sometimes it is better to have an approximation than to have no data at all. Compare Fig. 1-13 with Fig. 1-6, and you can see that the *linear interpolation error* is considerable in this case.

CURVE FITTING

Curve fitting is an intuitive scheme for approximating a point-to-point graph, or filling in a graph containing one or more gaps, to make it look like a continuous curve. Figure 1-14 is an approximate graph of the price of hypothetical Stock Y, based on points determined at intervals of half an hour, as generated by curve fitting. Here, the moment-to-moment stock price is shown by the dashed line, and the fitted curve, based on half-hour intervals, is shown by the solid line. The fitted curve does not precisely represent the actual stock price at every instant, but it comes close most of the time.

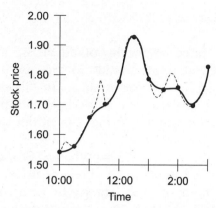

Fig. 1-14.   Approximation of hypothetical stock price as a continuous function of time, making use of curve fitting. The solid curve represents the approximation; the dashed curve represents the actual, moment-to-moment stock price as a function of time.

Curve fitting becomes increasingly accurate as the values are determined at more and more frequent intervals. When the values are determined infrequently, this scheme can be subject to large errors, as is shown by the example of Fig. 1-15.

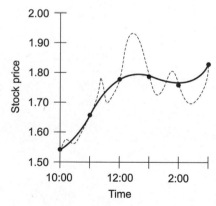

Fig. 1-15.   An example of curve fitting in which not enough data samples are taken, causing significant errors. The solid line represents the approximation; the dashed curve represents the actual, moment-to-moment stock price as a function of time.

EXTRAPOLATION

The term *extrapolate* means "to put outside of." When a function has a continuous-curve graph where time is the independent variable, *extrapolation* is the same thing as short-term forecasting. Two examples are shown in Fig. 1-16.

In Fig. 1-16A, the price of the hypothetical Stock X is plotted until 2:00 P.M., and then an attempt is made to forecast its price for an hour into the future, based on its past performance. In this case, *linear extrapolation*, the simplest form, is used. The curve is simply projected ahead as a straight line. Compare this graph with Fig. 1-6. In this case, linear extrapolation works fairly well.

Figure 1-16B shows the price of the hypothetical Stock Y plotted until 2:00 P.M., and then linear extrapolation is used in an attempt to predict its behavior for the next hour. As you can see by comparing this graph with Fig. 1-6, linear extrapolation does not work well in this scenario.

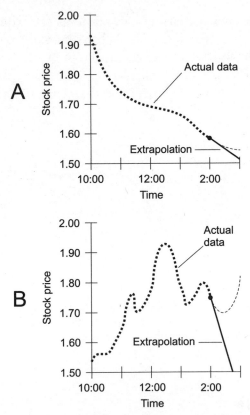

Fig. 1-16. Examples of linear extrapolation. The solid lines represent the forecasts; the dashed curves represent the actual data. In the case shown at A, the prediction is fairly good. In the case shown at B, the linear extrapolation is way off.

Extrapolation is best done by computers. Machines can notice subtle characteristics of functions that humans miss. Some graphs are easy to extrapolate, and others are not. In general, as a curve becomes more complicated, extrapolation becomes subject to more error. Also, as the extent (or distance) of the extrapolation increases for a given curve, the accuracy decreases.

TRENDS

A function is said to be *nonincreasing* if the value of the dependent variable never grows any larger (or more positive) as the value of the independent variable increases. If the dependent variable in a function never gets any smaller (or more negative) as the value of the independent variable increases, the function is said to be *nondecreasing*.

The dashed curve in Fig. 1-17 shows the behavior of a hypothetical Stock Q, whose price never rises throughout the period under consideration. This function is nonincreasing. The solid curve shows a hypothetical Stock R, whose price never falls throughout the period. This function is nondecreasing.

Sometimes the terms *trending downward* and *trending upward* are used to describe graphs. These terms are subjective; different people might interpret them differently. Everyone would agree that Stock Q in Fig. 1-17 trends downward while Stock R trends upward. But a stock that rises and falls several times during a period might be harder to define in this respect.

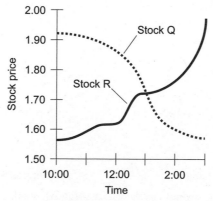

Fig. 1-17.   The price of Stock Q is nonincreasing versus time, and the price of Stock R is nondecreasing versus time.

CORRELATION

Specialized graphs called *scatter plots* or *scatter graphs* can show the extent of *correlation* between the values of two variables when the values are obtained from a finite number of experimental samples.

If, as the value of one variable generally increases, the value of the other generally increases too, the correlation is considered positive. If the opposite is true – the value of one variable generally increases as the other generally decreases – the correlation is considered negative. If the points are randomly scattered all over the graph, then the correlation is considered to be 0.

Figure 1-18 shows five examples of scatter plots. At A the correlation is 0. At B and C, the correlation is positive. At D and E, the correlation is

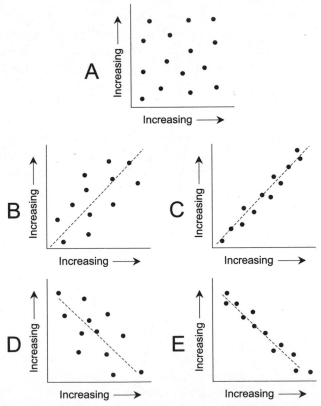

Fig. 1-18. Scatter plots showing correlation of 0 (at A), weak positive correlation (at B), strong positive correlation (at C), weak negative correlation (at D), and strong negative correlation (at E).

negative. When correlation exists, the points tend to be clustered along a well-defined path. In these examples the paths are straight lines, but in some situations they can be curves. The more nearly the points in a scatter plot lie along a straight line, the stronger the correlation.

Correlation is rated on a scale from a minimum of -1, through 0, up to a maximum of $+1$. When all the points in a scatter plot lie along a straight line that ramps downward as you go to the right, indicating that one variable decreases uniformly as the other variable increases, the correlation is -1. When all the points lie along a straight line that ramps upward as you go to the right, indicating that one variable increases uniformly as the other variable increases, the correlation is $+1$. None of the graphs in Fig. 1-18 show either of these extremes. The actual value of the correlation factor for a set of points is determined according to a rather complicated formula that is beyond the scope of this book.

PROBLEM 1-8

Suppose, as the value of the independent variable in a function changes, the value of the dependent variable does not change. This is called a *constant function*. Is its graph nonincreasing or nondecreasing?

SOLUTION 1-8

According to our definitions, the graph of a constant function is both non-increasing and nondecreasing. Its value never increases, and it never decreases.

PROBLEM 1-9

Is there any type of function for which linear interpolation is perfectly accurate, that is, "fills in the gap" with zero error?

SOLUTION 1-9

Yes. If the graph of a function is known to be a straight line, then linear interpolation can be used to "fill in a gap" and the result will be free of error. An example is the speed of a car that accelerates at a defined and constant rate. If its speed-versus-time graph appears as a perfectly straight line with a small gap, then linear interpolation can be used to determine the car's speed at points inside the gap, as shown in Fig. 1-19. In this graph, the heavy dashed line represents actual measured data, and the thinner solid line represents interpolated data.

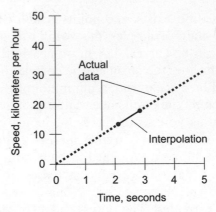

Fig. 1-19. Illustration for Problem 1-9.

Quiz

Refer to the text in this chapter if necessary. A good score is 8 correct. Answers are in the back of the book.

1. What is the square of $x/(-5)$, where x represents any real number? (The square of a number is equal to the number multiplied by itself.)
   (a) $x^2/5$
   (b) $x^2/25$
   (c) $-x^2/5$
   (d) $-x^2/25$

2. In the Venn diagram of Fig. 1-20, what does the shaded region represent?
   (a) $A + B$
   (b) $A \cap B$
   (c) $A \cup B$
   (d) $A \subseteq B$

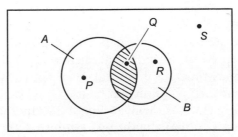

Fig. 1-20. Illustration for Quiz Questions 2 and 3.

3. In Fig. 1-20, the objects labeled *P*, *Q*, and *R* represent individual points. Which of the following statements is true?
   (a) $P \cap (A \in B)$
   (b) $Q \in (A \cap B)$
   (c) $R \in \varnothing$
   (d) $P \subset A$

4. When we write 23.723723723723 ..., we are writing a representation of
   (a) an irrational number
   (b) a nonterminating, nonrepeating decimal
   (c) a nondenumerable number
   (d) a rational number

5. Suppose a graph is drawn showing temperature in degrees Celsius (°C) as a function of time for a 24-hour day. The temperature measurements are accurate to within 0.1°C, and the readings are taken every 15 minutes. The day's highest temperature is +18.2°C at 4:45 P.M. The day's lowest temperature is +11.6°C at 5:00 A.M. A reasonable range for the temperature scale in this graph is
   (a) 0.0°C to +100.0°C
   (b) −100.0°C to +100.0°C
   (c) +10.0°C to +20.0°C
   (d) +10.0° C to +50.0°C

6. In a constant function:
   (a) the value of the dependent variable constantly increases as the value of the independent variable increases
   (b) the value of the dependent variable constantly decreases as the value of the independent variable increases
   (c) the value of the independent variable does not change
   (d) the value of the dependent variable does not change

7. It is reasonable to suppose that the number of lightning strikes per year in a 10,000-square mile region is positively correlated with
   (a) the number of people in the region
   (b) the number of stormy days per year in the region
   (c) the number of cars on the region's highways
   (d) the average altitude of the region above sea level

8. Suppose a large image file is downloaded from the Internet. The speed of the data, in bits per second (bps), is plotted as a function of time in seconds. In this situation, data speed is considered

(a) the dependent variable
(b) the independent variable
(c) a constant function
(d) nondecreasing

9. Suppose the path of a hurricane's center is plotted as a set of points on a map at 12-hour intervals, with the latitude and longitude lines as graphic coordinates. A continuous plot of the path of the storm center can be approximated using
(a) interpolation
(b) extrapolation
(c) curve fitting
(d) correlation

10. Which of the following statements is false?
(a) All relations are constant functions.
(b) All constant functions are relations.
(c) The values in a bar graph do not always add up to 100%.
(d) Zero correlation is indicated by widely scattered points on a graph.

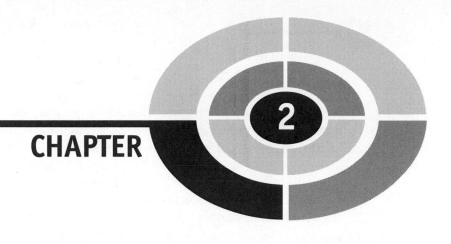

Learning the Jargon

Statistics is the analysis of information. In particular, statistics is concerned with *data*: information expressed as measurable or observable quantities. Statistical data is usually obtained by looking at the real world or universe, although it can also be generated by computers in "artificial worlds."

Experiments and Variables

If you want to understand anything about a scientific discipline, you must know the terminology. Statistics is no exception. Here are definitions of some terms used in statistics.

EXPERIMENT

In statistics, an *experiment* is an act of collecting data with the intent of learning or discovering something. For example, we might conduct an experiment to determine the most popular channels for frequency-modulation (FM) radio broadcast stations whose transmitters are located in American towns having less than 5000 people. Or we might conduct an experiment to determine the lowest barometric pressures inside the eyes of all the Atlantic hurricanes that take place during the next 10 years.

Experiments often, but not always, require specialized instruments to measure quantities. If we conduct an experiment to figure out the average test scores of high-school seniors in Wyoming who take a certain standardized test at the end of this school year, the only things we need are the time, energy, and willingness to collect the data. But a measurement of the minimum pressure inside the eye of a hurricane requires sophisticated hardware in addition to time, energy, and courage.

VARIABLE (IN GENERAL)

In mathematics, a *variable*, also called an *unknown*, is a quantity whose value is not necessarily specified, but that can be determined according to certain rules. Mathematical variables are expressed using italicized letters of the alphabet, usually in lowercase. For example, in the expression $x + y + z = 5$, the letters x, y, and z are variables that represent numbers.

In statistics, variables are similar to those in mathematics. But there are some subtle distinctions. Perhaps most important is this: In statistics, a variable is always associated with one or more experiments.

DISCRETE VARIABLE

In statistics, a *discrete variable* is a variable that can attain only specific values. The number of possible values is countable. Discrete variables are like the channels of a television set or digital broadcast receiver. It's easy to express the value of a discrete variable, because it can be assumed exact.

When a disc jockey says "This is radio 97.1," it means that the assigned channel center is at a frequency of 97.1 megahertz, where a megahertz (MHz) represents a million cycles per second. The assigned value is exact, even though, in real life, the broadcast engineer can at best get the transmitter output close to 97.1 MHz. The assigned channels in the FM broadcast band are separated by an *increment* (minimum difference) of 0.2 MHz. The next lower channel from 97.1 MHz is at 96.9 MHz, and the next higher one is at 97.3 MHz. There is no "in between." No two channels can be closer together than 0.2 MHz in the set of assigned standard FM broadcast channels in the

United States. The lowest channel is at 88.1 MHz and the highest is at 107.9 MHz (Fig. 2-1).

Other examples of discrete variables are:

- The number of people voting for each of the various candidates in a political election.
- The scores of students on a standardized test (expressed as a percentage of correct answers).
- The number of car drivers caught speeding every day in a certain town.
- The earned-run averages of pitchers in a baseball league (in runs per 9 innings or 27 outs).

All these quantities can be expressed as exact values. There is no error involved when discrete variables are measured or calculated.

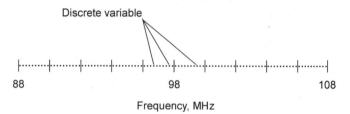

Fig. 2-1.   The individual channels in the FM broadcast band are values of a discrete variable.

CONTINUOUS VARIABLE

A *continuous variable* can attain infinitely many values over a certain span or range. Instead of existing as specific values in which there is an increment between any two, a continuous variable can change value to an arbitrarily tiny extent.

Continuous variables are something like the set of radio frequencies to which an analog FM broadcast receiver can be tuned. The radio frequency is adjustable continuously, say from 88 MHz to 108 MHz for an FM headset receiver with analog tuning (Fig. 2-2). If you move the tuning dial a little, you can make the received radio frequency change by something less than 0.2 MHz, the separation between adjacent assigned transmitter channels. There is no limit to how small the increment can get. If you have a light enough touch, you can adjust the received radio frequency by 0.02 MHz, or 0.002 MHz, or even 0.000002 MHz.

Other examples of continuous variables are:

- Temperature in degrees Celsius.

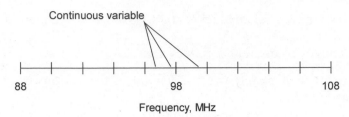

Fig. 2-2. The frequency to which an analog FM broadcast receiver can be set is an example of a continuous variable.

- Barometric pressure in millibars.
- Brightness of a light source in candela.
- Intensity of the sound from a loudspeaker in decibels with respect to the threshold of hearing.

Such quantities can never be determined exactly. There is always some instrument or observation error, even if that error is so small that it does not have a practical effect on the outcome of an experiment.

Populations and Samples

In statistics, the term *population* refers to a particular set of items, objects, phenomena, or people being analyzed. These items, also called *elements*, can be actual subjects such as people or animals, but they can also be numbers or definable quantities expressed in physical units.

Consistent with the above definitions of variables, some examples of populations are as follows:

- Assigned radio frequencies (in megahertz) of all FM broadcast transmitters in the United States.
- Temperature readings (in degrees Celsius) at hourly intervals last Wednesday at various locations around the city of New York.
- Minimum barometric-pressure levels (in millibars) at the centers of all the hurricanes in recorded history.
- Brightness levels (in candela) of all the light bulbs in offices in Minneapolis.
- Sound-intensity levels (in decibels relative to the threshold of hearing) of all the electric vacuum cleaners in the world.

SAMPLE, EVENT, AND CENSUS

A *sample* of a population is a subset of that population. It can be a set consisting of only one value, reading, or measurement singled out from a population, or it can be a subset that is identified according to certain characteristics. The physical unit (if any) that defines a sample is always the same as the physical unit that defines the main, or parent, population. A single element of a sample is called an *event*.

Consistent with the above definitions of variables, some samples are:

- Assigned radio frequencies of FM broadcast stations whose transmitters are located in the state of Ohio.
- Temperature readings at 1:00 P.M. local time last Wednesday at various locations around the city of New York.
- Minimum barometric-pressure levels at the centers of Atlantic hurricanes during the decade 1991–2000.
- Brightness levels of halogen bulbs in offices in Minneapolis.
- Sound-intensity levels of the electric vacuum cleaners used in all the households in Rochester, Minnesota.

When a sample consists of the whole population, it is called a *census*. When a sample consists of a subset of a population whose elements are chosen at random, it is called a *random sample*.

RANDOM VARIABLE

A *random variable* is a discrete or continuous variable whose value cannot be predicted in any given instance. Such a variable is usually defined within a certain range of values, such as 1 through 6 in the case of a thrown die, or from 88 MHz to 108 MHz in the case of an FM broadcast channel.

It is often possible to say, in a given scenario, that some values of a random variable are more likely to turn up than others. In the case of a thrown die, assuming the die is not "weighted," all of the values 1 through 6 are equally likely to turn up. When considering the FM broadcast channels of public radio stations, it is tempting to suppose (but this would have to be confirmed by observation) that transmissions are made more often at the lower radio-frequency range than at the higher range. Perhaps you have noticed that there is a greater concentration of public radio stations in the 4-MHz-wide sample from 88 MHz to 92 MHz than in, say, the equally wide sample from 100 MHz to 104 MHz.

In order for a variable to be random, the only requirement is that it be impossible to predict its value in any single instance. If you contemplate throwing a die one time, you can't predict how it will turn up. If you contemplate throwing a dart one time at a map of the United States while wearing a blindfold, you have no way of knowing, in advance, the lowest radio frequency of all the FM broadcast stations in the town nearest the point where the dart will hit.

FREQUENCY

The *frequency* of a particular outcome (result) of an event is the number of times that outcome occurs within a specific sample of a population. Don't confuse this with radio broadcast or computer processor frequencies! In statistics, the term "frequency" means "often-ness." There are two species of statistical frequency: *absolute frequency* and *relative frequency*.

Suppose you toss a die 6000 times. If the die is not "weighted," you should expect that the die will turn up showing one dot approximately 1000 times, two dots approximately 1000 times, and so on, up to six dots approximately 1000 times. The absolute frequency in such an experiment is therefore approximately 1000 for each face of the die. The relative frequency for each of the six faces is approximately 1 in 6, which is equivalent to about 16.67%.

PARAMETER

A specific, well-defined characteristic of a population is known as a *parameter* of that population. We might want to know such parameters as the following, concerning the populations mentioned above:

- The most popular assigned FM broadcast frequency in the United States.
- The highest temperature reading in the city of New York as determined at hourly intervals last Wednesday.
- The average minimum barometric-pressure level or measurement at the centers of all the hurricanes in recorded history.
- The lowest brightness level found in all the light bulbs in offices in Minneapolis.
- The highest sound-intensity level found in all the electric vacuum cleaners used in the world.

STATISTIC

A specific characteristic of a sample is called a *statistic* of that sample. We might want to know such statistics as these; concerning the samples mentioned above:

- The most popular assigned frequency for FM broadcast stations in Ohio.
- The highest temperature reading at 1:00 P.M. local time last Wednesday in New York.
- The average minimum barometric-pressure level or measurement at the centers of Atlantic hurricanes during the decade 1991–2000.
- The lowest brightness level found in all the halogen bulbs in offices in Minneapolis.
- The highest sound-intensity level found in electric vacuum cleaners used in households in Rochester, Minnesota.

Distributions

A *distribution* is a description of the set of possible values that a random variable can take. This can be done by noting the absolute or relative frequency. A distribution can be illustrated in terms of a table, or in terms of a graph.

DISCRETE VERSUS CONTINUOUS

Table 2-1 shows the results of a single, hypothetical experiment in which a die is tossed 6000 times. Figure 2-3 is a vertical bar graph showing the same data as Table 2-1. Both the table and the graph are distributions that describe the behavior of the die. If the experiment is repeated, the results will differ. If a huge number of experiments is carried out, assuming the die is not "weighted," the relative frequency of each face (number) turning up will approach 1 in 6, or approximately 16.67%.

Table 2-2 shows the number of days during the course of a 365-day year in which measurable precipitation occurs within the city limits of five different hypothetical towns. Figure 2-4 is a horizontal bar graph showing the same data as Table 2-2. Again, both the table and the graph are distributions. If the same experiment were carried out for several years in a row, the results would differ from year to year. Over a period of many years, the relative

Table 2-1 Results of a single, hypothetical experiment in which an "unweighted" die is tossed 6000 times.

Face of die	Number of times face turns up
1	968
2	1027
3	1018
4	996
5	1007
6	984

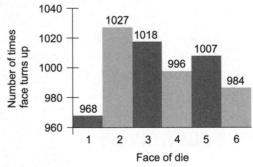

Fig. 2-3. Results of a single, hypothetical experiment in which an "unweighted" die is tossed 6000 times.

frequencies would converge towards certain values, although long-term climate change might have effects not predictable or knowable in our lifetimes.

Both of the preceding examples involve discrete variables. When a distribution is shown for a continuous variable, a graph must be used. Figure 2-5 is a distribution that denotes the relative amount of energy available from sunlight, per day during the course of a calendar year, at a hypothetical city in the northern hemisphere.

Table 2-2   Number of days on which measurable rain occurs in a specific year, in five hypothetical towns.

Town name	Number of days in year with measurable precipitation
Happyville	108
Joytown	86
Wonderdale	198
Sunnywater	259
Rainy Glen	18

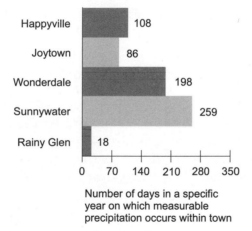

Fig. 2-4.   Measurable precipitation during a hypothetical year, in five different make-believe towns.

FREQUENCY DISTRIBUTION

In both of the above examples (the first showing the results of 6000 die tosses and the second showing the days with precipitation in five hypothetical towns), the scenarios are portrayed with frequency as the dependent variable. This is true of the tables as well as the graphs. Whenever frequency is portrayed as the dependent variable in a distribution, that distribution is called a *frequency distribution*.

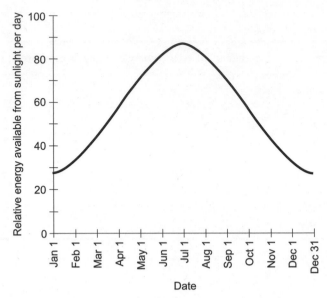

Fig. 2-5.   Relative energy available from sunlight, per day, during the course of a year at a hypothetical location.

Suppose we complicate the situation involving dice. Instead of one person tossing one die 6000 times, we have five people tossing five different dice, and each person tosses the same die 6000 times. The dice are colored red, orange, yellow, green, and blue, and are manufactured by five different companies, called Corp. A, Corp. B, Corp. C, Corp. D, and Corp. E, respectively. Four of the die are "weighted" and one is not. There are thus 30,000 die tosses to tabulate or graph in total. When we conduct this experiment, we can tabulate the data in at least two ways.

Ungrouped frequency distribution

The simplest way to tabulate the die toss results as a frequency distribution is to combine all the tosses and show the total frequency for each die face 1 through 6. A hypothetical example of this result, called an *ungrouped frequency distribution*, is shown in Table 2-3. We don't care about the weighting characteristics of each individual die, but only about potential biasing of the entire set. It appears that, for this particular set of die, there is some bias in favor of faces 4 and 6, some bias against faces 1 and 3, and little or no bias either for or against faces 2 and 5.

Table 2-3   An ungrouped frequency distribution showing the results of a single, hypothetical experiment in which five different die, some "weighted" and some not, are each tossed 6000 times.

Face of die	Toss results for all dice
1	4857
2	4999
3	4626
4	5362
5	4947
6	5209

Grouped frequency distribution

If we want to be more detailed, we can tabulate the frequency for each die face 1 through 6 separately for each die. A hypothetical product of this effort, called a *grouped frequency distribution*, is shown in Table 2-4. The results are grouped according to manufacturer and die color. From this distribution, it is apparent that some of the die are heavily "weighted." Only the green die, manufactured by Corp. D, seems to lack any bias. If you are astute, you will notice (or at least strongly suspect) that the green die here is the same die, with results gathered from the same experiment, as is portrayed in Table 2-1 and Fig. 2-3.

PROBLEM 2-1
Suppose you add up all the numbers in each column of Table 2-4. What should you expect, and why? What should you expect if the experiment is repeated many times?

SOLUTION 2-1
Each column should add up to 6000. This is the number of times each die (red, orange, yellow, green, or blue) is tossed in the experiment. If the sum of the numbers in any of the columns is not equal to 6000, then the experiment

Table 2-4   A grouped frequency distribution showing the results of a single, hypothetical experiment in which five different die, some "weighted" and some not, and manufactured by five different companies, are each tossed 6000 times.

Face of die	Toss results by manufacturer				
	Red Corp. A	Orange Corp. B	Yellow Corp. C	Green Corp. D	Blue Corp. E
1	625	1195	1689	968	380
2	903	1096	1705	1027	268
3	1300	890	1010	1018	408
4	1752	787	540	996	1287
5	577	1076	688	1007	1599
6	843	956	368	984	2058

was done in a faulty way, or else there is an error in the compilation of the table. If the experiment is repeated many times, the sums of the numbers in each column should always be 6000.

PROBLEM 2-2
Suppose you add up all the numbers in each row of Table 2-4. What should you expect, and why? What should you expect if the experiment is repeated many times?

SOLUTION 2-2
The sums of the numbers in the rows will vary, depending on the bias of the set of die considered as a whole. If, taken all together, the die show any bias, and if the experiment is repeated many times, the sums of the numbers should be consistently lower for some rows than for other rows.

PROBLEM 2-3
Each small rectangle in a table, representing the intersection of one row with one column, is called a *cell* of the table. What do the individual numbers in the cells of Table 2-4 represent?

SOLUTION 2-3
The individual numbers are absolute frequencies. They represent the actual number of times a particular face of a particular die came up during the course of the experiment.

More Definitions

Here are some more definitions you should learn in order to get comfortable reading or talking about statistics.

TRUNCATION

The process of *truncation* is a method of approximating numbers denoted as decimal expansions. It involves the deletion of all the numerals to the right of a certain point in the decimal part of an expression. Some electronic calculators use truncation to fit numbers within their displays. For example, the number 3.830175692803 can be shortened in steps as follows:

$$3.830175692803$$
$$3.83017569280$$
$$3.8301756928$$
$$3.830175692$$
$$3.83017569$$
$$3.8301756$$
$$3.830175$$
$$3.83017$$
$$3.83$$
$$3.8$$
$$3$$

ROUNDING

Rounding is the preferred method of approximating numbers denoted as decimal expansions. In this process, when a given digit (call it r) is deleted at the right-hand extreme of an expression, the digit q to its left (which becomes the new r after the old r is deleted) is not changed if $0 \leq r \leq 4$. If $5 \leq r \leq 9$, then q is increased by 1 ("rounded up"). Most electronic calculators use rounding rather than truncation. If rounding is used, the number 3.830175692803 can be shortened in steps as follows:

$$3.830175692803$$
$$3.83017569280$$
$$3.8301756928$$
$$3.830175693$$
$$3.83017569$$
$$3.8301757$$
$$3.830176$$
$$3.83018$$
$$3.8302$$
$$3.830$$
$$3.83$$
$$3.8$$
$$4$$

CUMULATIVE ABSOLUTE FREQUENCY

When data are tabulated, the absolute frequencies are often shown in one or more columns. Look at Table 2-5, for example. This shows the results of the tosses of the blue die in the experiment we looked at a while ago. The first column shows the number on the die face. The second column shows the absolute frequency for each face, or the number of times each face turned up during the experiment. The third column shows the *cumulative absolute frequency*, which is the sum of all the absolute frequency values in table cells at or above the given position.

The cumulative absolute frequency numbers in a table always ascend (increase) as you go down the column. The total cumulative absolute frequency should be equal to the sum of all the individual absolute frequency numbers. In this instance, it is 6000, the number of times the blue die was tossed.

Table 2-5   Results of an experiment in which a "weighted" die is tossed 6000 times, showing absolute frequencies and cumulative absolute frequencies.

Face of die	Absolute frequency	Cumulative absolute frequency
1	380	380
2	268	648
3	408	1056
4	1287	2343
5	1599	3942
6	2058	6000

CUMULATIVE RELATIVE FREQUENCY

Relative frequency values can be added up down the columns of a table, in exactly the same way as the absolute frequency values are added up. When this is done, the resulting values, usually expressed as percentages, show the *cumulative relative frequency*.

Examine Table 2-6. This is a more detailed analysis of what happened with the blue die in the above-mentioned experiment. The first, second, and fourth columns in Table 2-6 are identical with the first, second, and third columns in Table 2-5. The third column in Table 2-6 shows the percentage represented by each absolute frequency number. These percentages are obtained by dividing the number in the second column by 6000, the total number of tosses. The fifth column shows the cumulative relative frequency, which is the sum of all the relative frequency values in table cells at or above the given position.

The cumulative relative frequency percentages in a table, like the cumulative absolute frequency numbers, always ascend as you go down the column. The total cumulative relative frequency should be equal to 100%. In this sense, the cumulative relative frequency column in a table can serve as a *checksum*, helping to ensure that the entries have been tabulated correctly.

Table 2-6   Results of an experiment in which a "weighted" die is tossed 6000 times, showing absolute frequencies, relative frequencies, cumulative absolute frequencies, and cumulative relative frequencies.

Face of die	Absolute frequency	Relative frequency	Cumulative absolute frequency	Cumulative relative frequency
1	380	6.33%	380	6.33%
2	268	4.47%	648	10.80%
3	408	6.80%	1056	17.60%
4	1287	21.45%	2343	39.05%
5	1599	26.65%	3942	65.70%
6	2058	34.30%	6000	100.00%

MEAN

The *mean* for a discrete variable in a distribution is the mathematical average of all the values. If the variable is considered over the entire population, the average is called the *population mean*. If the variable is considered over a particular sample of a population, the average is called the *sample mean*. There can be only one population mean for a population, but there can be many different sample means. The mean is often denoted by the lowercase Greek letter mu, in italics (μ). Sometimes it is also denoted by an italicized lowercase English letter, usually x, with a bar (vinculum) over it.

Table 2-7 shows the results of a 10-question test, given to a class of 100 students. As you can see, every possible score is accounted for. There are some people who answered all 10 questions correctly; there are some who did not get a single answer right. In order to determine the mean score for the whole class on this test – that is, the population mean, called μ_p – we must add up the scores of each and every student, and then divide by 100. First, let's sum the products of the numbers in the first and second columns. This will give us 100 times the population mean:

Table 2-7 Scores on a 10-question test taken by 100 students.

Test score	Absolute frequency	Letter grade
10	5	A
9	6	A
8	19	B
7	17	B
6	18	C
5	11	C
4	6	D
3	4	D
2	4	F
1	7	F
0	3	F

$$(10 \times 5) + (9 \times 6) + (8 \times 19) + (7 \times 17) + (6 \times 18) + (5 \times 11) + (4 \times 6)$$
$$+ (3 \times 4) + (2 \times 4) + (1 \times 7) + (0 \times 3)$$
$$= 50 + 54 + 152 + 119 + 108 + 55 + 24 + 12 + 8 + 7 + 0$$
$$= 589$$

Dividing this by 100, the total number of test scores (one for each student who turns in a paper), we obtain $\mu_p = 589/100 = 5.89$.

The teacher in this class has assigned letter grades to each score. Students who scored 9 or 10 correct received grades of A; students who got scores of 7 or 8 received grades of B; those who got scores of 5 or 6 got grades of C; those who got scores of 3 or 4 got grades of D; those who got less than 3 correct answers received grades of F. The assignment of grades, informally known as the "curve," is a matter of teacher temperament and doubtless would seem arbitrary to the students who took this test. (Some people might consider the "curve" in this case to be overly lenient, while a few might think it is too severe.)

PROBLEM 2-4

What are the sample means for each grade in the test whose results are tabulated in Table 2-7? Use rounding to determine the answers to two decimal places.

SOLUTION 2-4

Let's call the sample means μ_{sa} for the grade of A, μ_{sb} for the grade of B, and so on down to μ_{sf} for the grade of F.

To calculate μ_{sa}, note that 5 students received scores of 10, while 6 students got scores of 9, both scores good enough for an A. This is a total of $5 + 6$, or 11, students getting the grade of A. Therefore:

$$\begin{aligned}
\mu_{sa} &= [(5 \times 10) + (6 \times 9)]/11 \\
&= (50 + 54)/11 \\
&= 104/11 \\
&= 9.45
\end{aligned}$$

To find μ_{sb}, observe that 19 students scored 8, and 17 students scored 7. Thus, $19 + 17$, or 36, students received grades of B. Calculating:

$$\begin{aligned}
\mu_{sb} &= [(19 \times 8) + (17 \times 7)]/36 \\
&= (152 + 119)/36 \\
&= 271/36 \\
&= 7.53
\end{aligned}$$

To determine μ_{sc}, check the table to see that 18 students scored 6, while 11 students scored 5. Therefore, $18 + 11$, or 29, students did well enough for a C. Grinding out the numbers yields this result:

$$\begin{aligned}
\mu_{sc} &= [(18 \times 6) + (11 \times 5)]/29 \\
&= (108 + 55)/29 \\
&= 163/29 \\
&= 5.62
\end{aligned}$$

To calculate μ_{sd}, note that 6 students scored 4, while 4 students scored 3. This means that $6 + 4$, or 10, students got grades of D:

$$\mu_{sd} = [(6 \times 4) + (4 \times 3)]/10$$
$$= (24 + 12)/10$$
$$= 36/10$$
$$= 3.60$$

Finally, we determine μ_{sf}. Observe that 4 students got scores of 2, 7 students got scores of 1, and 3 students got scores of 0. Thus, $4 + 7 + 3$, or 14, students failed the test:

$$\mu_{sf} = [(4 \times 2) + (7 \times 1) + (3 \times 0)]/14$$
$$= (8 + 7 + 0)/14$$
$$= 15/14$$
$$= 1.07$$

MEDIAN

If the number of elements in a distribution is even, then the *median* is the value such that half the elements have values greater than or equal to it, and half the elements have values less than or equal to it. If the number of elements is odd, then the median is the value such that the number of elements having values greater than or equal to it is the same as the number of elements having values less than or equal to it. The word "median" is synonymous with "middle."

Table 2-8 shows the results of the 10-question test described above, but instead of showing letter grades in the third column, the cumulative absolute frequency is shown instead. The tally is begun with the top-scoring papers and proceeds in order downward. (It could just as well be done the other way, starting with the lowest-scoring papers and proceeding upward.) When the scores of all 100 individual papers are tallied this way, so they are in order, the scores of the 50th and 51st papers – the two in the middle – are found to be 6 correct. Thus, the median score is 6, because half the students scored 6 or above, and the other half scored 6 or below.

It's possible that in another group of 100 students taking this same test, the 50th paper would have a score of 6 while the 51st paper would have a score of 5. When two values "compete," the median is equal to their average. In this case it would be midway between 5 and 6, or 5.5.

Table 2-8   The median can be determined by tabulating the cumulative absolute frequencies.

Test score	Absolute frequency	Cumulative absolute frequency
10	5	5
9	6	11
8	19	30
7	17	47
6 (partial)	3	50
6 (partial)	15	65
5	11	76
4	6	82
3	4	86
2	4	90
1	7	97
0	3	100

MODE

The *mode* for a discrete variable is the value that occurs the most often. In the test whose results are shown in Table 2-7, the most "popular" or often-occurring score is 8 correct answers. There were 19 papers with this score. No other score had that many results. Therefore, the mode in this case is 8.

Suppose that another group of students took this test, and there were two scores that occurred equally often. For example, suppose 16 students got 8 answers right, and 16 students also got 6 answers right. In this case there are two modes: 6 and 8. This sort of distribution is called a *bimodal distribution*.

Now imagine there are only 99 students in a class, and there are exactly 9 students who get each of the 11 possible scores (from 0 to 10 correct answers). In this distribution, there is no mode. Or, we might say, the mode is not defined.

The mean, median, and mode are sometimes called *measures of central tendency*. This is because they indicate a sort of "center of gravity" for the values in a data set.

VARIANCE

There is still another way in which the nature of a distribution can be described. This is a measure of the extent to which the values are spread out. There is something inherently different about a distribution of test scores like those portrayed in Table 2-7, compared with a distribution where every score is almost equally "popular." The test results portrayed in Table 2-7 are also qualitatively different than a distribution where almost every student got the same score, say 7 answers correct.

In the scenario of Table 2-7, call the variable x, and let the 100 individual scores be called x_1 through x_{100}. Suppose we find the extent to which each individual score x_i (where i is an integer between 1 and 100) differs from the mean score for the whole population (μ_p). This gives us 100 "distances from the mean," d_1 through d_{100}, as follows:

$$d_1 = |x_1 - \mu_p|$$
$$d_2 = |x_2 - \mu_p|$$
$$\downarrow$$
$$d_{100} = |x_{100} - \mu_p|$$

The vertical lines on each side of an expression represent the absolute value. For any real number r, $|r| = r$ if $r \geq 0$, and $|r| = -r$ if $r < 0$. The absolute value of a number is the extent to which it differs from 0. It avoids the occurrence of negative numbers as the result of a mathematical process.

Now, let's square each of these "distances from the mean," getting this set of numbers:

$$d_1{}^2 = (x_1 - \mu_p)^2$$
$$d_2{}^2 = (x_2 - \mu_p)^2$$
$$\downarrow$$
$$d_{100}{}^2 = (x_{100} - \mu_p)^2$$

The absolute-value signs are not needed in these expressions, because for any real number r, r^2 is never negative.

Next, let's average all the "squares of the distances from the mean," d_i^2. This means we add them all up, and then divide by 100, the total number of scores, obtaining the "average of the squares of the distances from the mean." This is called the *variance* of the variable x, written $\text{Var}(x)$:

$$\text{Var}(x) = (1/100)(d_1^2 + d_2^2 + \ldots + d_{100}^2)$$
$$= (1/100)[(x_1 - \mu_p)^2 + (x_2 - \mu_p)^2 + \ldots + (x_{100} - \mu_p)^2]$$

The variance of a set of n values whose population mean is μ_p is given by the following formula:

$$\text{Var}(x) = (1/n)[(x_1 - \mu_p)^2 + (x_2 - \mu_p)^2 + \ldots + (x_n - \mu_p)^2]$$

STANDARD DEVIATION

Standard deviation, like variance, is an expression of the extent to which values are spread out with respect to the mean. The standard deviation is the square root of the variance, and is symbolized by the italicized, lowercase Greek letter sigma (σ). (Conversely, the variance is equal to the square of the standard deviation, and is often symbolized σ^2.) In the scenario of our test:

$$\sigma = [(1/100)(d_1^2 + d_2^2 + \ldots + d_{100}^2)]^{1/2}$$
$$= \{(1/100)[(x_1 - \mu_p)^2 + (x_2 - \mu_p)^2 + \ldots + (x_{100} - \mu_p)^2]\}^{1/2}$$

The standard deviation of a set of n values whose population mean is μ_p is given by the following formula:

$$\sigma = \{(1/n)[(x_1 - \mu_p)^2 + (x_2 - \mu_p)^2 + \ldots + (x_n - \mu_p)^2]\}^{1/2}$$

These expressions are a little messy. It's easier for some people to remember these verbal definitions:

- Variance is the average of the squares of the "distances" of each value from the mean.
- Standard deviation is the square root of the variance.

Variance and standard deviation are sometimes called *measures of dispersion*. In this sense the term "dispersion" means "spread-outedness."

PROBLEM 2-5
Draw a vertical bar graph showing all the absolute-frequency data from Table 2-5, the results of a "weighted" die-tossing experiment. Portray each die face on the horizontal axis. Let light gray vertical bars show the absolute

frequency numbers, and let dark gray vertical bars show the cumulative absolute frequency numbers.

SOLUTION 2-5
Figure 2-6 shows such a graph. The numerical data is not listed at the tops of the bars in order to avoid excessive clutter.

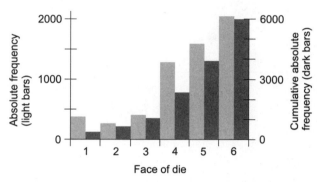

Fig. 2-6.   Illustration for Problem 2-5.

PROBLEM 2-6
Draw a horizontal bar graph showing all the relative-frequency data from Table 2-6, another portrayal of the results of a "weighted" die-tossing experiment. Show each die face on the vertical axis. Let light gray horizontal bars show the relative frequency percentages, and dark gray horizontal bars show the cumulative relative frequency percentages.

SOLUTION 2-6
Figure 2-7 is an example of such a graph. Again, the numerical data is not listed at the ends of the bars, in the interest of neatness.

PROBLEM 2-7
Draw a point-to-point graph showing the absolute frequencies of the 10-question test described by Table 2-7. Mark the population mean, the median, and the mode with distinctive vertical lines, and label them.

SOLUTION 2-7
Figure 2-8 is an example of such a graph. Numerical data is included for the population mean, median, and mode.

PROBLEM 2-8
Calculate the variance, Var(x), for the 100 test scores tabulated in Table 2-7.

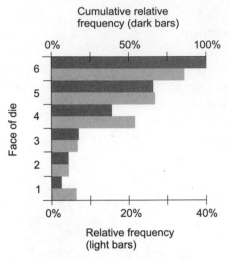

Fig. 2-7. Illustration for Problem 2-6.

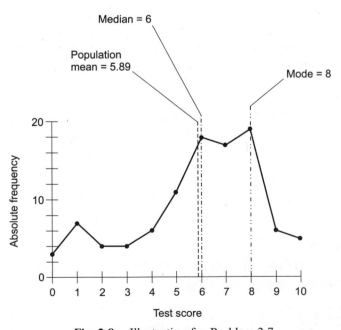

Fig. 2-8. Illustration for Problem 2-7.

SOLUTION 2-8

Recall that the population mean, μ_p, as determined above, is 5.89. Table 2-9 shows the "distances" of each score from μ_p, the squares of these "distances," and the products of each of these squares with the absolute frequencies (the number of papers having each score from 0 to 10). At the bottom of the table, these products are all summed. The resulting number, 643.58, is 100 times the variance. Therefore:

$$\text{Var}(x) = (1/100)(643.58)$$
$$= 6.4358$$

It is reasonable to round this off to 6.44.

Table 2-9   "Distances" of each test score x_i from the population mean μ_p, the squares of these "distances," the products of the squares with the absolute frequencies f_i, and the sum of these products. This information is used in the solution of Problem 2-8.

Test score	Abs. freq. f_i	"Distance" from mean: $x_i - \mu_p$	"Distance" squared: $(x_i - \mu_p)^2$	Product: $f_i \times (x_i - \mu_p)^2$
10	5	+4.11	16.89	84.45
9	6	+3.11	9.67	58.02
8	19	+2.11	4.45	84.55
7	17	+1.11	1.23	20.91
6	18	+0.11	0.01	0.18
5	11	−0.89	0.79	8.69
4	6	−1.89	3.57	21.42
3	4	−2.89	8.35	33.40
2	4	−3.89	15.13	60.52
1	7	−4.89	23.91	167.37
0	3	−5.89	34.69	104.07
Sum of products $f_i \times (x_i - \mu_p)^2$				643.58

PROBLEM 2-9
Calculate the standard deviation for the 100 test scores tabulated in Table 2-7.

SOLUTION 2-9
The standard deviation, σ, is the square root of the variance. Approximating:

$$\sigma = [\text{Var}(x)]^{1/2}$$
$$= 6.4358^{1/2}$$
$$= 2.5369$$

It is reasonable to round this off to 2.54.

Quiz

Refer to the text in this chapter if necessary. A good score is 8 correct. Answers are in the back of the book.

1. Suppose a large number of people take a test, and every single student gets exactly half of the answers right. In this case, the standard deviation is
    (a) equal to the mean
    (b) equal to the median
    (c) equal to zero
    (d) impossible to determine without more information

2. In a frequency distribution:
    (a) the frequency is the dependent variable
    (b) the median is always equal to the mean
    (c) the mode represents the average value
    (d) the mean is always equal to the mode

3. A tabulation of cumulative absolute frequency values is useful in determining
    (a) the mode
    (b) the variance
    (c) the median
    (d) the mean

4. A subset of a population is known as
    (a) a sample

(b) a continuous variable
(c) a random variable
(d) a discrete variable

5. Imagine that 11 people take a 10-question test. Suppose one student gets 10 correct answers, one gets 9 correct, one gets 8 correct, and so on, all the way down to one student getting none correct. What is the mean score, accurate to three decimal places?
(a) 4.545
(b) 5.000
(c) 5.500
(d) It is not defined.

6. Imagine that 11 people take a 10-question test. Suppose one student gets 10 correct answers, one gets 9 correct, one gets 8 correct, and so on, all the way down to one student getting none correct. What is the median score, accurate to three decimal places?
(a) 4.545
(b) 5.000
(c) 5.500
(d) It is not defined.

7. Imagine that 11 people take a 10-question test. Suppose one student gets 10 correct answers, one gets 9 correct, one gets 8 correct, and so on, all the way down to one student getting none correct. What is the mode score, accurate to three decimal places?
(a) 4.545
(b) 5.000
(c) 5.500
(d) It is not defined.

8. Suppose a variable λ (lambda, pronounced "LAM-da," a lowercase Greek letter commonly used in physics) can attain a value equal to any positive real number. In this instance, λ is an example of
(a) a continuous variable
(b) a discrete variable
(c) an absolute variable
(d) a relative variable

9. The largest cumulative absolute frequency in a set of numbers is equal to
(a) the sum of all the individual absolute frequency values
(b) twice the mean

(c) twice the median

(d) twice the mode

10. Which of the following is an expression of the extent to which values are spread out relative to the mean?

(a) The average.

(b) The mode.

(c) The median.

(d) None of the above.

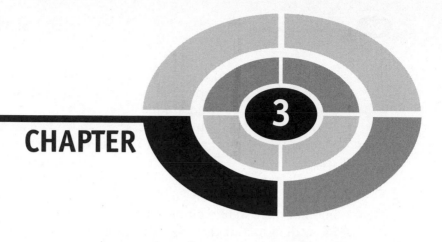

Basics of Probability

Probability is the proportion or percentage of the time that specified things happen. The term *probability* is also used in reference to the art and science of determining the proportion or percentage of the time that specified things happen.

The Probability Fallacy

We say something is true because we've seen or deduced it. If we believe something is true or has taken place but we aren't sure, it's tempting to say it is or was "likely." It's wise to resist this temptation.

BELIEF

When people formulate a theory, they often say that something "probably" happened in the distant past, or that something "might" exist somewhere, as-yet undiscovered, at this moment. Have you ever heard that there is a "good chance" that extraterrestrial life exists? Such a statement is meaningless. Either it exists, or it does not.

If you say "I believe the universe began with an explosion," you are stating the fact that you believe it, not the fact that it is true or that it is "probably" true. If you say "The universe began with an explosion!" your statement is logically sound, but it is a statement of a theory, not a proven fact. If you say "The universe probably started with an explosion," you are in effect suggesting that there were multiple pasts and the universe had an explosive origin in more than half of them. This is an instance of what can be called the *probability fallacy* (abbreviated PF), wherein probability is injected into a discussion inappropriately.

Whatever is, is. Whatever is not, is not. Whatever was, was. Whatever was not, was not. Either the universe started with an explosion, or it didn't. Either there is life on some other world, or there isn't.

PARALLEL WORLDS?

If we say that the "probability" of life existing elsewhere in the cosmos is 20%, we are in effect saying, "Out of n observed universes, where n is some large number, $0.2n$ universes have been found to have extraterrestrial life." That doesn't mean anything to those of us who have seen only one universe!

It is worthy of note that there are theories involving so-called *fuzzy truth*, in which some things "sort of happen." These theories involve degrees of truth that span a range over which probabilities can be assigned to occurrences in the past and present. An example of this is *quantum mechanics*, which is concerned with the behavior of subatomic particles. Quantum mechanics can get so bizarre that some scientists say, "If you claim to understand this stuff, then you are lying." We aren't going to deal with anything that esoteric.

WE MUST OBSERVE

Probability is usually defined according to the results of observations, although it is sometimes defined on the basis of theory alone. When the notion of probability is abused, seemingly sound reasoning can be employed to come to absurd conclusions. This sort of thing is done in industry every day, especially when the intent is to get somebody to do something that will cause somebody else to make money. Keep your "probability fallacy radar" on when navigating through the real world.

If you come across an instance where an author says that something "probably happened," "is probably true," "is likely to take place," or "is not likely to happen," think of it as another way of saying that the author believes or suspects that something did or didn't happen, is or isn't true, or is or is not expected to take place on the basis of experimentation or observation. I can tell you right now that I'm probably going to make statements later in this book to which this clarification should be applied. Maybe I've already done it!

Key Definitions

Here are definitions of some common terms that will help us understand what we are talking about when we refer to probability.

EVENT VERSUS OUTCOME

The terms *event* and *outcome* are easily confused. An event is a single occurrence or trial in the course of an experiment. An outcome is the result of an event.

If you toss a coin 100 times, there are 100 separate events. Each event is a single toss of the coin. If you throw a pair of dice simultaneously 50 times, each act of throwing the pair is an event, so there are 50 events.

Suppose, in the process of tossing coins, you assign "heads" a value of 1 and "tails" a value of 0. Then when you toss a coin and it comes up "heads," you can say that the outcome of that event is 1. If you throw a pair of dice and get a sum total of 7, then the outcome of that event is 7.

The outcome of an event depends on the nature of the hardware and processes involved in the experiment. The use of a pair of "weighted" dice produces different outcomes, for an identical set of events, than a pair of "unweighted" dice. The outcome of an event also depends on how the event is defined. There is a difference between saying that the sum is 7 in a toss of two dice, as compared with saying that one of the dice comes up 2 while the other one comes up 5.

SAMPLE SPACE

A *sample space* is the set of all possible outcomes in the course of an experiment. Even if the number of events is small, a sample space can be large.

If you toss a coin four times, there are 16 possible outcomes. These are listed in Table 3-1, where "heads" = 1 and "tails" = 0. (If the coin happens to land on its edge, you disregard that result and toss it again.)

Table 3-1   The sample space for an experiment in which a coin is tossed four times. There are 16 possible outcomes; "heads" = 1 and "tails" = 0.

Event 1	Event 2	Event 3	Event 4
0	0	0	0
0	0	0	1
0	0	1	0
0	0	1	1
0	1	0	0
0	1	0	1
0	1	1	0
0	1	1	1
1	0	0	0
1	0	0	1
1	0	1	0
1	0	1	1
1	1	0	0
1	1	0	1
1	1	1	0
1	1	1	1

If a pair of dice, one red and one blue, is tossed once, there are 36 possible outcomes in the sample space, as shown in Table 3-2. The outcomes are denoted as ordered pairs, with the face-number of the red die listed first and the face-number of the blue die listed second.

Table 3-2   The sample space for an experiment consisting of a single event, in which a pair of dice (one red, one blue) is tossed once. There are 36 possible outcomes, shown as ordered pairs (red,blue).

Red → Blue ↓	1	2	3	4	5	6
1	(1,1)	(2,1)	(3,1)	(4,1)	(5,1)	(6,1)
2	(1,2)	(2,2)	(3,2)	(4,2)	(5,2)	(6,2)
3	(1,3)	(2,3)	(3,3)	(4,3)	(5,3)	(6,3)
4	(1,4)	(2,4)	(3,4)	(4,4)	(5,4)	(6,4)
5	(1,5)	(2,5)	(3,5)	(4,5)	(5,5)	(6,5)
6	(1,6)	(2,6)	(3,6)	(4,6)	(5,6)	(6,6)

MATHEMATICAL PROBABILITY

Let x be a discrete random variable that can attain n possible values, all equally likely. Suppose an outcome H results from exactly m different values of x, where $m \leq n$. Then the *mathematical probability* $p_{math}(H)$ that outcome H will result from any given value of x is given by the following formula:

$$p_{math}(H) = m/n$$

Expressed as a percentage, the probability $p_{\%}(H)$ is:

$$p_{math\%}(H) = 100m/n$$

If we toss an "unweighted" die once, each of the six faces is as likely to turn up as each of the others. That is, we are as likely to see 1 as we are to see 2, 3, 4, 5, or 6. In this case, there are six possible values, so $n=6$. The mathematical probability of any particular face turning up ($m=1$) is equal to $p_{math}(H)=1/6$. To calculate the mathematical probability of either of any two different faces turning up (say 3 or 5), we set $m = 2$; therefore $p_{math}(H) = 2/6 = 1/3$. If we want to know the mathematical probability that any one of the six faces will turn up, we set $m = 6$, so the formula gives us $p_{math}(H) = 6/6 = 1$. The respective percentages $p_{math\%}(H)$ in these cases are 16.67% (approximately), 33.33% (approximately), and 100% (exactly).

Mathematical probabilities can only exist within the range 0 to 1 (or 0% to 100%) inclusive. The following formulas describe this constraint:

$$0 \leq p_{math}(H) \leq 1$$
$$0\% \leq p_{math\%}(H) \leq 100\%$$

We can never have a mathematical probability of 2, or −45%, or −6, or 556%. When you give this some thought, it is obvious. There is no way for something to happen less often than never. It's also impossible for something to happen more often than all the time.

EMPIRICAL PROBABILITY

In order to determine the likelihood that an event will have a certain outcome in real life, we must rely on the results of prior experiments. The probability of a particular outcome taking place, based on experience or observation, is called *empirical probability*.

Suppose we are told that a die is "unweighted." How does the person who tells us this know that it is true? If we want to use this die in some application, such as when we need an object that can help us to generate a string of random numbers from the set {1, 2, 3, 4, 5, 6}, we can't take on faith the notion that the die is "unweighted." We have to check it out. We can analyze the die in a lab and figure out where its center of gravity is; we measure how deep the indentations are where the dots on its faces are inked. We can scan the die electronically, X-ray it, and submerge it in (or float it on) water. But to be absolutely certain that the die is "unweighted," we must toss it many thousands of times, and be sure that each face turns up, on the average, 1/6 of the time. We must conduct an experiment – gather *empirical evidence* – that supports the contention that the die is "unweighted." Empirical probability is based on determinations of relative frequency, which was discussed in the last chapter.

As with mathematical probability, there are limits to the range an empirical probability figure can attain. If H is an outcome for a particular single event, and the empirical probability of H taking place as a result of that event is denoted $p_{emp}(H)$, then:

$$0 \leq p_{\mathrm{emp}}(H) \leq 1$$
$$0\% \leq p_{\mathrm{emp}\%}(H) \leq 100\%$$

PROBLEM 3-1
Suppose a new cholesterol-lowering drug comes on the market. If the drug is to be approved by the government for public use, it must be shown effective, and it must also be shown not to have too many serious side effects. So it is tested. During the course of testing, 10,000 people, all of whom have been diagnosed with high cholesterol, are given this drug. Imagine that 7289 of the people experience a significant drop in cholesterol. Also suppose that 307 of these people experience adverse side effects. If you have high cholesterol and go on this drug, what is the empirical probability $p_{\mathrm{emp}}(B)$ that you will derive benefit? What is the empirical probability $p_{\mathrm{emp}}(A)$ that you will experience adverse side effects?

SOLUTION 3-1
Some readers will say that this question cannot be satisfactorily answered because the experiment is not good enough. Is 10,000 test subjects a large enough number? What physiological factors affect the way the drug works? How about blood type, for example? Ethnicity? Gender? Blood pressure? Diet? What constitutes "high cholesterol"? What constitutes a "significant drop" in cholesterol level? What is an "adverse side effect"? What is the standard drug dose? How long must the drug be taken in order to know if it works? For convenience, we ignore all of these factors here, even though, in a true scientific experiment, it would be an excellent idea to take them all into consideration.

   Based on the above experimental data, shallow as it is, the relative frequency of effectiveness is 7289/10,000 = 0.7289 = 72.89%. The relative frequency of ill effects is 307/10,000 = 0.0307 = 3.07%. We can round these off to 73% and 3%. These are the empirical probabilities that you will derive benefit, or experience adverse effects, if you take this drug in the hope of lowering your high cholesterol. Of course, once you actually use the drug, these probabilities will lose all their meaning for you. You will eventually say "The drug worked for me" or "The drug did not work for me." You will say, "I had bad side effects" or "I did not have bad side effects."

REAL-WORLD EMPIRICISM

Empirical probability is used by scientists to make predictions. It is not good for looking at aspects of the past or present. If you try to calculate the empirical probability of the existence of extraterrestrial life in our galaxy, you can play around with formulas based on expert opinions, but once you state a numeric figure, you commit the PF. If you say the empirical probability that a hurricane of category 3 or stronger struck the U.S. mainland in 1992 equals $x\%$ (where $x < 100$) because at least one hurricane of that intensity hit the U.S. mainland in x of the years in the 20th century, historians will tell you that is rubbish, as will anyone who was in Homestead, Florida on August 24, 1992.

Imperfection is inevitable in the real world. We can't observe an infinite number of people and take into account every possible factor in a drug test. We cannot toss a die an infinite number of times. The best we can hope for is an empirical probability figure that gets closer and closer to the "absolute truth" as we conduct a better and better experiment. Nothing we can conclude about the future is a "totally sure bet."

Properties of Outcomes

Here are some formulas that describe properties of outcomes in various types of situations. Don't let the symbology intimidate you. It is all based on the set theory notation covered in Chapter 1.

LAW OF LARGE NUMBERS

Suppose you toss an "unweighted" die many times. You get numbers turning up, apparently at random, from the set $\{1, 2, 3, 4, 5, 6\}$. What will the average value be? For example, if you toss the die 100 times, total up the numbers on the faces, and then divide by 100, what will you get? Call this number d (for die). It is reasonable to suppose that d will be fairly close to the mean, μ:

$$d \approx \mu$$
$$d \approx (1 + 2 + 3 + 4 + 5 + 6)/6$$
$$= 21/6$$
$$= 3.5$$

It's possible, in fact likely, that if you toss a die 100 times you'll get a value of d that is slightly more or less than 3.5. This is to be expected because of "reality imperfection." But now imagine tossing the die 1000 times, or 100,000 times, or even 100,000,000 times! The "reality imperfections" will be smoothed out by the fact that the number of tosses is so huge. The value of d will converge to 3.5. As the number of tosses increases without limit, the value of d will get closer and closer to 3.5, because the opportunity for repeated coincidences biasing the result will get smaller and smaller.

The foregoing scenario is an example of the *law of large numbers*. In a general, informal way, it can be stated like this: "As the number of events in an experiment increases, the average value of the outcome approaches the theoretical mean." This is one of the most important laws in all of probability theory.

INDEPENDENT OUTCOMES

Two outcomes H_1 and H_2 are *independent* if and only if the occurrence of one does not affect the probability that the other will occur. We write it this way:

$$p(H_1 \cap H_2) = p(H_1)\, p(H_2)$$

Figure 3-1 illustrates this situation in the form of a Venn diagram. The intersection is shown by the darkly shaded region.

A good example of independent outcomes is the tossing of a penny and a nickel. The face ("heads" or "tails") that turns up on the penny has no effect on the face ("heads" or "tails") that turns up on the nickel. It does not matter whether the two coins are tossed at the same time or at different times. They never interact with each other.

To illustrate how the above formula works in this situation, let $p(P)$ represent the probability that the penny turns up "heads" when a penny and a nickel are both tossed once. Clearly, $p(P) = 0.5$ (1 in 2). Let $p(N)$ represent the probability that the nickel turns up "heads" in the same scenario. It's obvious that $p(N) = 0.5$ (also 1 in 2). The probability that both coins turn up "heads" is, as you should be able to guess, 1 in 4, or 0.25. The above formula states it this way, where the intersection symbol $\cap$ can be translated as "and":

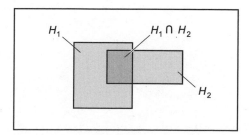

Fig. 3-1. Venn diagram showing intersection.

$$p(P \cap N) = p(P)p(N)$$
$$= 0.5 \times 0.5$$
$$= 0.25$$

MUTUALLY EXCLUSIVE OUTCOMES

Let H_1 and H_2 be two outcomes that are *mutually exclusive*; that is, they have no elements in common:

$$H_1 \cap H_2 = \varnothing$$

In this type of situation, the probability of either outcome occurring is equal to the sum of their individual probabilities. Here's how we write it, with the union symbol $\cup$ translated as "either/or":

$$p(H_1 \cup H_2) = p(H_1) + p(H_2)$$

Figure 3-2 shows this as a Venn diagram.

When two outcomes are mutually exclusive, they cannot both occur. A good example is the tossing of a single coin. It's impossible for "heads" and "tails" to both turn up on a given toss. But the sum of the two probabilities (0.5 for "heads" and 0.5 for "tails" if the coin is "balanced") is equal to the probability (1) that one or the other outcome will take place.

Another example is the result of a properly run, uncomplicated election for a political office between two candidates. Let's call the candidates Mrs. Anderson and Mr. Boyd. If Mrs. Anderson wins, we get outcome A, and if Mr. Boyd wins, we get outcome B. Let's call the respective probabilities of their winning $p(A)$ and $p(B)$. We might argue about the actual values of $p(A)$ and $p(B)$. We might obtain empirical probability figures by conducting a poll prior to the election, and get the idea that $p_{\text{emp}}(A) = 0.29$ and $p_{\text{emp}}(B) = 0.71$. We can be certain, however, of two facts: the two candidates

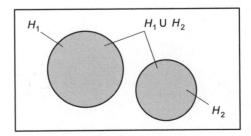

Fig. 3-2.   Venn diagram showing a pair of mutually exclusive outcomes.

won't both win, but one of them will. The probability that either Mrs. Anderson or Mr. Boyd will win is equal to the sum of $p(A)$ and $p(B)$, whatever these values happen to be, and we can be sure it is equal to 1 (assuming neither of the candidates quits during the election and is replaced by a third, unknown person, and assuming there are no write-ins or other election irregularities). Mathematically:

$$p(A \cup B) = p(A) + p(B)$$
$$= p_{emp}(A) + p_{emp}(B)$$
$$= 0.29 + 0.71$$
$$= 1$$

COMPLEMENTARY OUTCOMES

If two outcomes H_1 and H_2 are *complementary*, then the probability, expressed as a ratio, of one outcome is equal to 1 minus the probability, expressed as a ratio, of the other outcome. The following equations hold:

$$p(H_2) = 1 - p(H_1)$$
$$p(H_1) = 1 - p(H_2)$$

Expressed as percentages:

$$p_\%(H_2) = 100 - p_\%(H_1)$$
$$p_\%(H_1) = 100 - p_\%(H_2)$$

Figure 3-3 shows this as a Venn diagram.

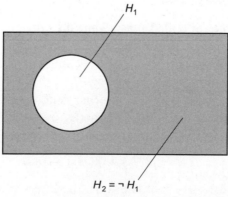

H_1

$H_2 = \neg H_1$

Fig. 3-3.   Venn diagram showing a pair of complementary outcomes.

The notion of complementary outcomes is useful when we want to find the probability that an outcome will fail to occur. Consider again the election between Mrs. Anderson and Mr. Boyd. Imagine that you are one of those peculiar voters who call themselves "contrarians," and who vote against, rather than for, candidates in elections. You are interested in the probability that "your candidate" (the one you dislike more) will lose. According to the pre-election poll, $p_{emp}(A) = 0.29$ and $p_{emp}(B) = 0.71$. We might state this inside-out as:

$$p_{emp}(\neg B) = 1 - p_{emp}(B)$$
$$= 1 - 0.71$$
$$= 0.29$$

$$p_{emp}(\neg A) = 1 - p_{emp}(A)$$
$$= 1 - 0.29$$
$$= 0.71$$

where the "droopy minus sign" ($\neg$) stands for the "not" operation, also called *logical negation*. If you are fervently wishing for Mr. Boyd to lose, then you can guess from the poll that the likelihood of your being happy after the election is equal to $p_{emp}(\neg B)$, which is 0.29 in this case.

Note that in order for two outcomes to be complementary, the sum of their probabilities must be equal to 1. This means that one or the other (but not both) of the two outcomes must take place; they are the only two possible outcomes in a scenario.

NONDISJOINT OUTCOMES

Outcomes H_1 and H_2 are called *nondisjoint* if and only if they have at least one element in common:

$$H_1 \cap H_2 \neq \varnothing$$

In this sort of case, the probability of either outcome is equal to the sum of the probabilities of their occurring separately, minus the probability of their occurring simultaneously. The equation looks like this:

$$p(H_1 \cup H_2) = p(H_1) + p(H_2) - p(H_1 \cap H_2)$$

Figure 3-4 shows this as a Venn diagram. The intersection of probabilities is subtracted in order to ensure that the elements common to both sets (repre-

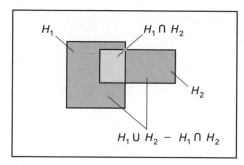

Fig. 3-4. Venn diagram showing a pair of nondisjoint outcomes.

sented by the lightly shaded region where the two sets overlap) are counted only once.

PROBLEM 3-2

Imagine that a certain high school has 1000 students. The new swimming and diving coach, during his first day on the job, is looking for team prospects. Suppose that the following are true:

- 200 students can swim well enough to make the swimming team
- 100 students can dive well enough to make the diving team
- 30 students can make either team or both teams

If the coach wanders through the hallways blindfolded and picks a student at random, determine the probabilities, expressed as ratios, that the coach will pick

- a fast swimmer; call this $p(S)$
- a good diver; call this $p(D)$
- someone good at both swimming and diving; call this $p(S \cap D)$
- someone good at either swimming or diving, or both; call this $p(S \cup D)$

SOLUTION 3-2

This problem is a little tricky. We assume that the coach has objective criteria for evaluating prospective candidates for his teams! That having been said, we must note that the outcomes are not mutually exclusive, nor are they independent. There is overlap, and there is interaction. We can find the first three answers immediately, because we are told the numbers:

$$p(S) = 200/1000 = 0.200$$
$$p(D) = 100/1000 = 0.100$$
$$p(S \cap D) = 30/1000 = 0.030$$

In order to calculate the last answer – the total number of students who can make either team or both teams – we must find $p(S \cup D)$ using this formula:

$$p(S \cup D) = p(S) + p(D) - p(S \cap D)$$
$$= 0.200 + 0.100 - 0.030$$

$$= 0.270$$

This means that 270 of the students in the school are potential candidates for either or both teams. The answer is not 300, as one might at first expect. That would be the case only if there were no students good enough to make both teams. We mustn't count the exceptional students twice. (However well somebody can act like a porpoise, he or she is nevertheless only one person!)

MULTIPLE OUTCOMES

The formulas for determining the probabilities of mutually exclusive and nondisjoint outcomes can be extended to situations in which there are three possible outcomes.

Three mutually exclusive outcomes. Let H_1, H_2, and H_3 be three mutually exclusive outcomes, such that the following facts hold:

$$H_1 \cap H_2 = \varnothing$$
$$H_1 \cap H_3 = \varnothing$$
$$H_2 \cap H_3 = \varnothing$$

The probability of any one of the three outcomes occurring is equal to the sum of their individual probabilities (Fig. 3-5):

$$p(H_1 \cup H_2 \cup H_3) = p(H_1) + p(H_2) + p(H_3)$$

Three nondisjoint outcomes. Let H_1, H_2, and H_3 be three nondisjoint outcomes. This means that one or more of the following facts is true:

$$H_1 \cap H_2 \neq \varnothing$$
$$H_1 \cap H_3 \neq \varnothing$$
$$H_2 \cap H_3 \neq \varnothing$$

The probability of any one of the outcomes occurring is equal to the sum of the probabilities of their occurring separately, minus the probabilities of each pair occurring simultaneously, minus the probability of all three occurring simultaneously (Fig. 3-6):

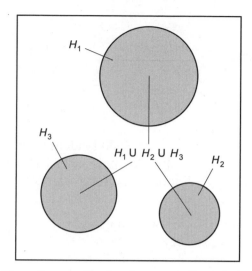

Fig. 3-5. Venn diagram showing three mutually exclusive outcomes.

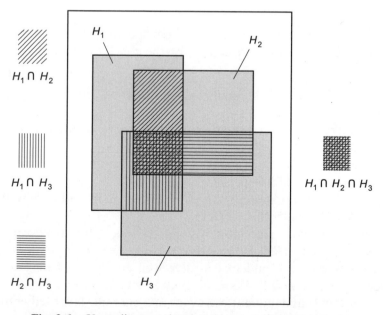

Fig. 3-6. Venn diagram showing three nondisjoint outcomes.

$$p(H_1 \cup H_2 \cup H_3)$$
$$= p(H_1) + p(H_2) + p(H_3)$$
$$- p(H_1 \cap H_2) - p(H_1 \cap H_3) - p(H_2 \cap H_3)$$
$$- p(H_1 \cap H_2 \cap H_3)$$

PROBLEM 3-3

Consider again the high school with 1000 students. The coach seeks people for the swimming, diving, and water polo teams in the same wandering, blindfolded way as before. Suppose the following is true of the students in the school:

- 200 people can make the swimming team
- 100 people can make the diving team
- 150 people can make the water polo team
- 30 people can make both the swimming and diving teams
- 110 people can make both the swimming and water polo teams
- 20 people can make both the diving and water polo teams
- 10 people can make all three teams

If the coach staggers around and tags students at random, what is the probability, expressed as a ratio, that the coach will, on any one tag, select a student who is good enough for at least one of the sports?

SOLUTION 3-3

Let the following expressions stand for the respective probabilities, all representing the results of random selections by the coach (and all of which we are told):

- Probability that a student can swim fast enough $= p(S) = 200/1000 = 0.200$.
- Probability that a student can dive well enough $= p(D) = 100/1000 = 0.100$.
- Probability that a student can play water polo well enough $= p(W) = 150/1000 = 0.150$.
- Probability that a student can swim fast enough and dive well enough $= p(S \cap D) = 30/1000 = 0.030$.
- Probability that a student can swim fast enough and play water polo well enough $= p(S \cap W) = 110/1000 = 0.110$.
- Probability that a student can dive well enough and play water polo well enough $= p(D \cap W) = 20/1000 = 0.020$.
- Probability that a student can swim fast enough, dive well enough, and play water polo well enough $= p(S \cap D \cap W) = 10/1000 = 0.010$.

In order to calculate the total number of students who can end up playing at least one sport for this coach, we must find $p(S \cup D \cup W)$ using this formula:

$$p(S \cup D \cup W) = p(S) + p(D) + p(W)$$
$$- p(S \cap D) - p(S \cap W) - p(D \cap W)$$
$$- p(S \cap D \cap W)$$
$$= 0.200 + 0.100 + 0.150$$
$$- 0.030 - 0.110 - 0.020 - 0.010$$
$$= 0.280$$

This means that 280 of the students in the school are potential prospects.

Permutations and Combinations

In probability, it is often necessary to choose small sets from large ones, or to figure out the number of ways in which certain sets of outcomes can take place. *Permutations* and *combinations* are the two most common ways this is done. Before we can define these, however, we need to define a function of non-negative integers called the *factorial*.

FACTORIAL

The factorial of a number is indicated by writing an exclamation point after it. If n is a natural number and $n \geq 1$, the value of $n!$ is defined as the product of all natural numbers less than or equal to n:

$$n! = 1 \times 2 \times 3 \times 4 \times \ldots \times n$$

If $n = 0$, then by default, $n! = 1$. The factorial is not defined for negative numbers.

As n increases, the value of $n!$ goes up rapidly, and when n reaches significant values, the factorial skyrockets. There is a formula for approximating $n!$ when n is large:

$$n! \approx n^n \, / \, e^n$$

where e is a constant called the *natural logarithm base*, and is equal to approximately 2.71828. The squiggly equals sign emphasizes the fact that the value of $n!$ using this formula is approximate, not exact.

PROBLEM 3-4
Write down the values of the factorial function for $n = 0$ through $n = 15$, in order to illustrate just how fast this value "blows up."

SOLUTION 3-4
The results are shown in Table 3-3. It's perfectly all right to use a calculator here. It should be capable of displaying a lot of digits. Most personal computers have calculators that are good enough for this purpose.

PROBLEM 3-5
Determine the approximate value of 100! using the formula given above.

SOLUTION 3-5
A calculator is not an option here; it is a requirement. You should use one that has an e^x (or natural exponential) function key. In case your calculator does not have this key, the value of the exponential function can be found by using the natural logarithm key and the inverse function key together. It will also be necessary for the calculator to have an x^y key (also called $x\hat{}y$) that lets you find the value of a number raised to its own power. In addition, the calculator should be capable of displaying numbers in *scientific notation*, also called *power-of-10 notation*. Most personal computer calculators are adequate if they are set for scientific mode.

   Using the above formula for $n = 100$:

$$100! \approx (100^{100}) / e^{100}$$
$$\approx (1.00 \times 10^{200}) / (2.688117 \times 10^{43})$$
$$\approx 3.72 \times 10^{156}$$

The numeral representing this number, if written out in full, would be a string of digits too long to fit on most text pages without taking up two or more lines. Your calculator will probably display it as something like 3.72e +156 or 3.72 E 156. In these displays, the "e" or "E" does not refer to the natural logarithm base. Instead, it means "times 10 raised to the power of."

PERMUTATIONS

Suppose q and r are both positive integers. Imagine a set of q items taken r at a time in a specific order. The possible number of permutations in this situation is symbolized $_qP_r$ and can be calculated as follows:

$$_qP_r = q! / (q - r)!$$

Table 3-3   Values of $n!$ for $n = 0$ through $n = 15$. This table constitutes the solution to Problem 3-4.

Value of n	Value of $n!$
0	0
1	1
2	2
3	6
4	24
5	120
6	720
7	5040
8	40,320
9	362,880
10	3,628,800
11	39,916,800
12	479,001,600
13	6,227,020,800
14	87,178,291,200
15	1,307,674,368,000

COMBINATIONS

Suppose q and r are positive integers. Imagine a set of q items taken r at a time in no particular order. The possible number of combinations in this situation is symbolized $_qC_r$ and can be calculated as follows:

$$_qC_r = {_qP_r} / r! = q! / [r!(q - r)!]$$

PROBLEM 3-6

How many permutations are there if you have 10 apples, taken 5 at a time in a specific order?

SOLUTION 3-6

Use the above formula for permutations, plugging in $q = 10$ and $r = 5$:

$$_{10}P_5 = 10! / (10 - 5)!$$
$$= 10! / 5!$$
$$= 10 \times 9 \times 8 \times 7 \times 6$$
$$= 30{,}240$$

PROBLEM 3-7

How many combinations are there if you have 10 apples, taken 5 at a time in no particular order?

SOLUTION 3-7

Use the above formula for combinations, plugging in $q = 10$ and $r = 5$. We can use the formula that derives combinations based on permutations, because we already know from the previous problem that $_{10}P_5 = 30{,}240$:

$$_{10}C_5 = {_{10}P_5} / 5!$$
$$= 30{,}240 / 120$$
$$= 252$$

The Density Function

When working with large populations, and especially with continuous random variables, probabilities are defined differently than they are with small populations and discrete random variables. As the number of possible values of a random variable becomes larger and "approaches infinity," it's easier to think of the probability of an outcome within a range of values, rather than the probability of an outcome for a single value.

A PATTERN EMERGES

Imagine that some medical researchers want to find out how people's blood pressure levels compare. At first, a few dozen people are selected at random from the human population, and the numbers of people having each of 10 specific systolic pressure readings are plotted (Fig. 3-7A). The systolic pressure, which is the higher of the two numbers you get when you take your blood pressure, is the random variable. (In this example, exact numbers

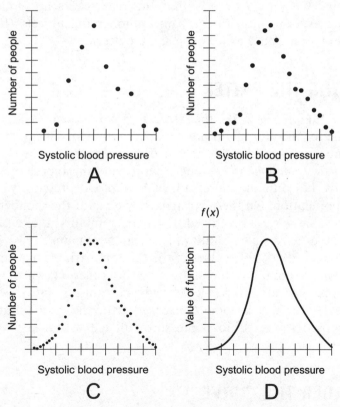

Fig. 3-7.   Hypothetical plots of blood pressure. At A, a small population and 10 values of pressure; at B, a large population and 20 values of pressure; at C, a gigantic population and 40 values of pressure; at D, the value of the density function versus blood pressure.

aren't shown either for blood pressure or for the number of people. This helps us keep in mind that this entire scenario is make-believe.)

There seems to be a pattern in Fig. 3-7A. This does not come as a surprise to our group of medical research scientists. They expect most people to have "middling" blood pressure, fewer people to have moderately low or high pressure, and only a small number of people to have extremely low or high blood pressure.

In the next phase of the experiment, hundreds of people are tested. Instead of only 10 different pressure levels, 20 discrete readings are specified for the random variable (Fig. 3-7B). A pattern is obvious. Confident that they're onto something significant, the researchers test thousands of people and plot the results at 40 different blood pressure levels. The resulting plot of frequency (number of people) versus the value of the random variable (blood pressure) shows that there is a highly defined pattern. The arrangement of points in Fig. 3-7C is so orderly that the researchers are confident that repeating the experiment with the same number of subjects (but not the same people) will produce exactly the same pattern.

EXPRESSING THE PATTERN

Based on the data in Fig. 3-7C, the researchers can use curve fitting to derive a general rule for the way blood pressure is distributed. This is shown in Fig. 3-7D. A smooth curve like this is called a *probability density function*, or simply a *density function*. It no longer represents the blood pressure levels of individuals, but only an expression of how blood pressure varies among the human population. On the vertical axis, instead of the number of people, the function value, $f(x)$, is portrayed. Does this remind you of a Cheshire cat that gradually dissolves away until only its smile remains?

As the number of possible values of the random variable increases without limit, the point-by-point plot blurs into a density function, which we call $f(x)$. The blood pressure of any particular subject vanishes into insignificance. Instead, the researchers become concerned with the probability that any randomly chosen subject's blood pressure will fall within a given range of values.

AREA UNDER THE CURVE

Figure 3-8 is an expanded view of the curve derived by "refining the points" of Fig. 3-7 to their limit. This density function, like all density functions, has a special property: if you calculate or measure the total area under the curve,

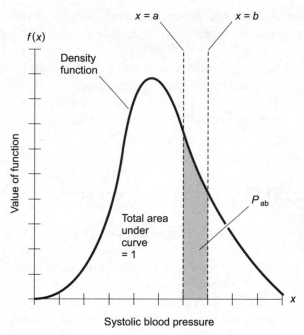

Fig. 3-8.   The probability of a randomly chosen value k of a random variable x falling between two limit values a and b is equal to the area under the curve between the vertical lines $x = a$ and $x = b$.

it is always equal to 1. This rule holds true for the same reason that the relative frequency values of the outcomes for a discrete variable always add up to 1 (or 100%), as we learned in the last chapter.

Consider two hypothetical systolic blood pressure values: say a and b as shown in Fig. 3-8. (Again, we refrain from giving specific numbers because this example is meant to be for illustration, not to represent any recorded fact.) The probability that a randomly chosen person will have a systolic blood pressure reading k that is between a and b can be written in any of four ways:

$$P(a < k < b)$$
$$P(a \leq k < b)$$
$$P(a < k \leq b)$$
$$P(a \leq k \leq b)$$

The first of these expressions includes neither a nor b, the second includes a but not b, the third includes b but not a, and the fourth includes both a and b. All four expressions are identical in the sense that they are all represented by the shaded portion of the area under the curve. Because of this, an expression with less-than signs only is generally used when expressing discrete probability.

If the vertical lines $x = a$ and $x = b$ are moved around, the area of the shaded region gets larger or smaller. This area can never be less than 0 (when the two lines coincide) or greater than 1 (when the two lines are so far apart that they allow for the entire area under the curve).

Two Common Distributions

The nature of a probability density function can be described in the form of a *distribution*. Let's get a little bit acquainted with two classical types: the *uniform distribution* and the *normal distribution*. These are mere introductions, intended to let you get to know what sorts of animals probability distributions are. There are many types of distributions besides these two, and each one of them could easily become the subject of an entire book. Distributions are sort of like spiders. In one sense, "when you've seen one, you've seen them all." But if you are willing to go deep, you can look at any species endlessly and keep discovering new things about it.

UNIFORM DISTRIBUTION

In a uniform distribution, the value of the function is constant. When graphed, it looks "flat," like a horizontal line (Fig. 3-9).

Let x be a continuous random variable. Let x_{min} and x_{max} be the minimum and maximum values that x can attain, respectively. In a uniform distribution, x has a density function of the form:

$$f(x) = 1 / (x_{max} - x_{min})$$

Because the total area under the curve is equal to 1, the probability P_{ab} that any randomly chosen x will be between a and b is:

$$P_{ab} = (b - a) / (x_{max} - x_{min})$$

Suppose that the experiment described above reveals that equal numbers of people always have each of the given tested blood-pressure numbers between two limiting values, say $x_{min} = 100$ and $x_{max} = 140$. Imagine that this is true no matter how many people are tested, and no matter how many different values of blood pressure are specified within the range $x_{min} = 100$

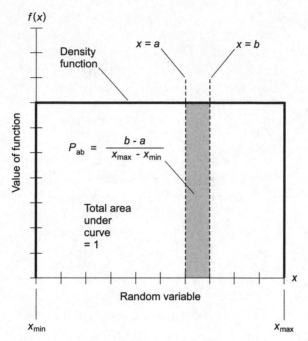

Fig. 3-9. A uniform density function has a constant value when the random variable is between two extremes x_{min} and x_{max}.

and $x_{max} = 140$. This is far-fetched and can't represent the real world, but if you play along with the idea, you can see that such a state of affairs produces a uniform probability distribution.

The mean (μ), the variance (σ^2), and the standard deviation (σ), which we looked at in the last chapter for discrete random variables, can all be defined for a uniform distribution having a continuous random variable. A detailed analysis of this is beyond the scope of this introductory course. But here are the formulas, in case you're interested:

$$\mu = (a + b) \,/\, 2$$
$$\sigma^2 = (b - a)^2 \,/\, 12$$
$$\sigma = [(b - a)^2 \,/\, 12]^{1/2}$$

NORMAL DISTRIBUTION

In a normal distribution, the value of the function has a single central peak, and tapers off on either side in a symmetrical fashion. Because of its shape, a graph of this function is often called a *bell-shaped curve* (Fig. 3-10). The

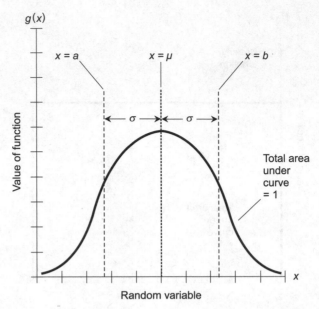

Fig. 3-10.   The normal distribution is also known as the bell-shaped curve.

normal distribution isn't just any bell-shaped curve, however. In order to be a true normal distribution, the curve must conform to specific rules concerning its standard deviation.

The symbol σ represents the standard deviation of the function, which is an expression of the extent to which the values of the function are concentrated. It's the same concept you saw in Chapter 2, but generalized for continuous random variables. A small value of σ produces a "sharp" curve with a narrow peak and steep sides. A large value of σ produces a "broad" curve with less steep sides. As σ approaches 0, the curve becomes narrower and narrower, closing in on a vertical line. If σ becomes arbitrarily large, the curve becomes almost flat and settles down near the horizontal axis. In any normal distribution, the area under the curve is equal to 1, no matter how much or little it is concentrated around the mean.

The symbol μ represents the mean, or average. Again, this is the same mean you learned about in Chapter 2, but generalized for continuous random variables. The value of μ can be found by imagining a moving vertical line that intersects the x axis. When the position of the vertical line is such that the area under the curve to its left is 1/2 (or 50%) and the area under the curve to its right is also 1/2 (50%), then the vertical line intersects the x axis at the point $x = \mu$. In the normal distribution, $x = \mu$ at the same point where the function attains its *peak*, or maximum, value.

THE EMPIRICAL RULE

Imagine two movable vertical lines, one on either side of the vertical line $x = \mu$. Suppose these vertical lines, $x = a$ and $x = b$, are such that the one on the left is the same distance from $x = \mu$ as the one on the right. The proportion of data points in the part of the distribution $a < x < b$ is defined by the proportion of the area under the curve between the two movable lines $x = a$ and $x = b$. Figure 3-10 illustrates this situation. A well-known theorem in statistics, called the *empirical rule*, states that all normal distributions have the following three characteristics:

- Approximately 68% of the data points are within the range $\pm\sigma$ of μ.
- Approximately 95% of the data points are within the range $\pm2\sigma$ of μ.
- Approximately 99.7% of the data points are within the range $\pm3\sigma$ of μ.

PROBLEM 3-8

Suppose you want it to rain so your garden will grow. It's a gloomy morning. The weather forecasters, who are a little bit weird in your town, expect a 50% chance that you'll see up to 1 centimeter (1 cm) of rain in the next 24 hours, and a 50% chance that more than 1 cm of rain will fall. They say it is impossible for more than 2 cm to fall (a dubious notion at best), and it is also impossible for less than 0 cm to fall (an absolute certainty!). Suppose the radio disc jockeys (DJs), who are even weirder than the meteorologists, announce the forecast and start talking about a distribution function called $R(x)$ for the rain as predicted by the weather experts. One DJ says that the amount of rain represents a continuous random variable x, and the distribution function $R(x)$ for the precipitation scenario is a normal distribution whose value tails off to 0 at precipitation levels of 0 cm and 2 cm. Draw a crude graph of what they're talking about.

SOLUTION 3-8

See Fig. 3-11. The portion of the curve to the left of the vertical line, which represents the mean, has an area of 0.5. The mean itself is $x = \mu = 1$ cm.

PROBLEM 3-9

Imagine the scenario above, as the DJs continue to expound. They start talking about the extent to which the distribution function is spread out around the mean value of 1 cm. One of them mentions that there is something called standard deviation, symbolized by a lowercase Greek letter called sigma that looks like a numeral 6 that has fallen over. Another DJ says that 68% of the total prospective future wetness of the town falls within

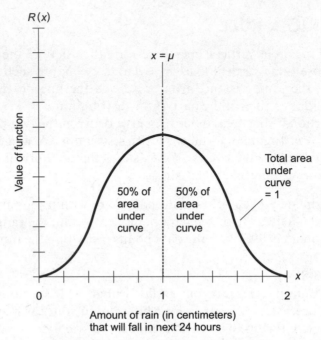

R(x)

x = μ

Value of function

Total area
under
curve
= 1

50% of
area
under
curve

50% of
area
under
curve

0   1   2   x

Amount of rain (in centimeters)
that will fall in next 24 hours

Fig. 3-11. Illustration for Problem 3-8.

two values of precipitation defined by sigma on either side of the mean. Draw
a crude graph of what they're talking about now.

SOLUTION 3-9
See Fig. 3-12. The shaded region represents the area under the curve between
the two vertical lines represented by $x = \mu - \sigma$ and $x = \mu + \sigma$. This is 68%
of the total area under the curve, centered at the vertical line representing the
mean, $x = \mu$.

PROBLEM 3-10
What does the standard deviation appear to be, based on the graph of Fig.
3-12?

SOLUTION 3-10
It looks like approximately $\sigma = 0.45$, representing the distance of either the
vertical line $x = \mu - \sigma$ or the vertical line $x = \mu + \sigma$ from the mean,
$x = \mu = 1$ cm. Note that this is only the result of the crude graph we've
drawn here. The DJs have not said anything about the actual value of σ.
We want to stick around and find out if they'll tell us what it is, but then there

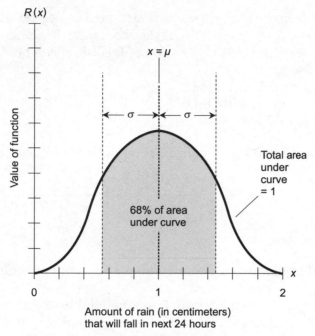

Fig. 3-12. Illustration for Problems 3-9 and 3-10.

is a massive lightning strike on the broadcast tower, and the station is knocked off the air.

Quiz

Refer to the text in this chapter if necessary. A good score is 8 correct. Answers are in the back of the book.

1. Empirical probability is based on
   (a) observation or experimentation
   (b) theoretical models only
   (c) continuous outcomes
   (d) standard deviations

2. Imagine a perfectly balanced spinning wheel, in which the likelihood of the pointer coming to rest within any particular range of directions (shaped like a thin slice of pie) is the same as the likelihood of its coming to rest within any other range of equal size. If the direction in

which the pointer comes to rest, measured in degrees of the compass (clockwise from true north), is the random variable, the density function that represents the behavior of this wheel can be described as
(a) a discrete distribution
(b) a normal distribution
(c) a uniform distribution
(d) none of the above

3. What is the number of possible permutations of 7 objects taken 3 at a time?
(a) 10
(b) 21
(c) 35
(d) 210

4. The difference between permutations and combinations lies in the fact that
(a) permutations take order into account, but combinations do not
(b) combinations take order into account, but permutations do not
(c) combinations involve only continuous variables, but permutations involve only discrete variables
(d) permutations involve only continuous variables, but combinations involve only discrete variables

5. The result of an event is called
(a) an experiment
(b) a trial
(c) an outcome
(d) a variable

6. Suppose some means is devised that quantitatively describes the political views of people on a continuum from "liberal" to "conservative," where the most "liberal" possible views are represented by a value of −50 (the extreme left end of the scale) and the most "conservative" possible views are represented by a value of +50 (the extreme right end of the scale). Suppose a density function is graphed and it is discovered, not surprisingly, that most people tend to fall near the "middle" and that the curve is bell-shaped. This represents
(a) a discrete distribution
(b) a normal distribution
(c) a uniform distribution
(d) none of the above

7. The set of all possible outcomes during the course of an experiment is called
   (a) a dependent variable
   (b) a random variable
   (c) a discrete variable
   (d) a sample space

8. What is the mathematical probability that a coin, tossed 10 times in a row, will come up "tails" on all 10 tosses?
   (a) 1
   (b) 1/10
   (c) 1/1024
   (d) 1/4096

9. Two outcomes are mutually exclusive if and only if
   (a) they are nondisjoint
   (b) they have no elements in common
   (c) they have at least one element in common
   (d) they have identical sets of outcomes

10. The probability, expressed as a percentage, of a particular occurrence can never be
    (a) less than 100
    (b) less than 0
    (c) greater than 1
    (d) anything but a whole number

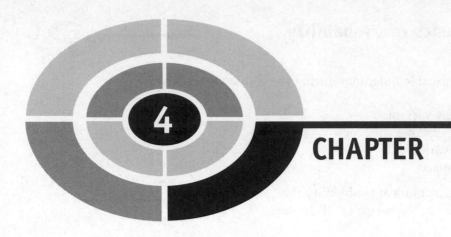

CHAPTER

Descriptive Measures

When analyzing data, it can be useful to break it down into intervals, graph it in special ways, or describe its characteristics in terms of numbers derived from certain formulas. These techniques are called *descriptive measures*.

Percentiles

When you were in elementary school, do you remember taking standardized tests every fall? In my time (the 1960s) and place (Minnesota) they were known as the Iowa Tests of Basic Skills. I recall being told at the end of one year's ordeal that I was in a certain *percentile*.

PERCENTILES IN A NORMAL DISTRIBUTION

Percentiles divide a large data set into 100 intervals, each interval containing 1% of the elements in the set. There are 99 possible percentiles, not 100, because the percentiles represent the boundaries where the 100 intervals meet.

Imagine an experiment in which systolic blood pressure readings are taken for a large number of people. The systolic reading is the higher of the two numbers the instrument displays. So if your reading is 110/70, read "110 over 70," the systolic pressure is 110. Suppose the results of this experiment are given to us in graphical form, and the curve looks like a continuous distribution because there the population is huge. Suppose it happens to be a normal distribution: bell-shaped and symmetrical (Fig. 4-1).

Let's choose some pressure value on the horizontal axis, and extend a line L straight up from it. The percentile corresponding to this pressure is determined by finding the number n such that at least $n\%$ of the area under the curve falls to the left of line L. Then, n is rounded to the nearest whole number between, and including, 1 and 99 to get the percentile p. For example, suppose the region to the left of the line L represents 93.3% of the area under the curve. Therefore, $n = 93.3$. Rounding to the nearest whole number between 1 and 99 gives $p = 93$. This means the blood pressure corresponding to the point where line L intersects the horizontal axis is in the 93rd percentile.

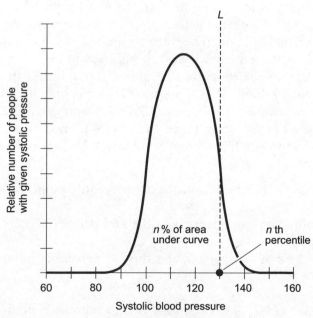

Fig. 4-1.   Determination of percentiles in a normal distribution.

The location of any particular percentile point (boundary), say the pth, is found by locating the vertical line such that the percentage n of the area beneath the curve is exactly equal to p, and then noting the point where this line crosses the horizontal axis. In Fig. 4-1, imagine that line L can be moved back and forth like a sliding door. When the number n, representing the percentage of the area beneath the curve to the left of L, is exactly equal to 93 then the line crosses the horizontal axis at the 93rd percentile boundary point. Although it's tempting to think that there could be a "0th percentile" ($n = 0$) and a "100th percentile" ($n = 100$), neither of these "percentiles" represents a boundary where two intervals meet.

Note the difference between saying that a certain pressure "is in" the pth percentile, versus saying that a certain pressure "is at" the pth percentile. In the first case we're describing a data interval; in the second case we're talking about a boundary point between two intervals.

PERCENTILES IN TABULAR DATA

Imagine that 1000 students take a 40-question test. There are 41 possible scores: 0 through 40. Suppose that every score is accounted for. There are some people who write perfect papers, and there are a few unfortunates who don't get any of the answers right. Table 4-1 shows the test results, with the scores in ascending order from 0 to 40 in the first column. For each possible score, the number of students getting that score (the absolute frequency) is shown in the second column. The third column shows the cumulative absolute frequency, expressed from lowest to highest scores.

Where do we put the 99 percentile points (boundaries) in this data set? How can we put 99 "fault lines" into a set that has only 41 possible values? The answer is, obviously, that we can't. What about grouping the students, then? A thousand people have taken the test. Why not break them up into 100 different groups with 99 different boundaries, and then call the 99 boundaries the "percentile points," like this?

- The "worst" 10 papers, and the 1st percentile point at the top of that group.
- The "2nd worst" 10 papers, and the 2nd percentile point at the top of that group.
- The "3rd worst" 10 papers, and the 3rd percentile point at the top of that group.
  ↓
- The "pth worst" 10 papers, and the pth percentile point at the top of that group.

Table 4-1 Results of a hypothetical 40-question test taken by 1000 students.

Test score	Absolute frequency	Cumulative absolute frequency
0	5	5
1	5	10
2	10	20
3	14	34
4	16	50
5	16	66
6	18	84
7	16	100
8	12	112
9	17	129
10	16	145
11	16	161
12	17	178
13	22	200
14	13	213
15	19	232
16	18	250
17	25	275
18	25	300
19	27	327

continued

Table 4.1 continued

20	33	360
21	40	400
22	35	435
23	30	465
24	35	500
25	31	531
26	34	565
27	35	600
28	34	634
29	33	667
30	33	700
31	50	750
32	50	800
33	45	845
34	27	872
35	28	900
36	30	930
37	28	958
38	20	978
39	12	990
40	10	1000

↓
- The "qth best" 10 papers, and the $(100 - q)$th percentile point at the top of that group.

↓
- The "3rd best" 10 papers, and the 97th percentile point at the bottom of that group.
- The "2nd best" 10 papers, and the 98th percentile point at the bottom of that group.
- The "best" 10 papers, and the 99th percentile point at the bottom of that group.

This looks great at first, but there's a problem. When we check Table 4-1, we can see that 50 people have scored 31 on the test. That's five groups of 10 people, all with the same score. These scores are all "equally good" (or "equally bad"). If we're going to say that any one of these papers is "in the pth percentile," then clearly we must say that they are all "in the pth percentile." We cannot arbitrarily take 10 papers with scores of 31 and put them in the pth percentile, then take 10 more papers with scores of 31 and put them in the $p + 1$st percentile, then take 10 more papers with scores of 31 and put them in the $p + 2$nd percentile, then take 10 more papers with scores of 31 and put them in the $p + 3$rd percentile, and then take 10 more papers with scores of 31 and put them in the $p + 4$th percentile. That would be unfair.

This business of percentiles is starting to get confusing and messy, isn't it? By now you must be wondering, "Who invented this concept, anyway?" That doesn't matter; the scheme is commonly used and we are stuck with it. What can we do to clear it all up and find a formula that makes sense in all possible scenarios?

PERCENTILE POINTS

We get around the foregoing conundrum by defining a scheme for calculating the positions of the percentile points in a set of *ranked data elements*. A set of ranked data elements is a set arranged in a table from "worst to best," such as that in Table 4-1. Once we have defined the percentile positioning scheme, we accept it as a convention, ending all confusion forever and ever.

So – suppose we are given the task of finding the position of the pth percentile in a set of n ranked data elements. First, multiply p by n, and then divide the product by 100. This gives us a number i called the *index*:

$$i = pn/100$$

Here are the rules:

- If i is not a whole number, then the location of the pth percentile point is $i + 1$.
- If i is a whole number, then the location of the pth percentile point is $i + 0.5$.

PERCENTILE RANKS

If we want to find the percentile rank p for a given element or position s in a ranked data set, we use a different definition. We divide the number of elements less than s (call this number t) by the total number of elements n, and multiply this quantity by 100, getting a tentative percentile p^*:

$$p^* = 100 \ t/n$$

Then we round p^* to the nearest whole number between, and including, 1 and 99 to get the percentile rank p for that element or position in the set.

   Percentile ranks defined in this way are intervals whose centers are at the percentile boundaries as defined above. The 1st and 99th percentile ranks are often a little bit oversized according to this scheme, especially if the population is large. This is because the 1st and 99th percentile ranks encompass *outliers*, which are elements at the very extremes of a set or distribution.

PERCENTILE INVERSION

Once in a while you'll hear people use the term "percentile" in an inverted, or upside-down, sense. They'll talk about the "first percentile" when they really mean the 99th, the "second percentile" when they really mean the 98th, and so on. Beware of this! If you get done with a test and think you have done well, and then you're told that you're in the "4th percentile," don't panic. Ask the teacher or test administrator, "What does that mean, exactly? The top 4%? The top 3%? The top 3.5%? Or what?" Don't be surprised if the teacher or test administrator is not certain.

PROBLEM 4-1
Where is the 56th percentile point in the data set shown by Table 4-1?

SOLUTION 4-1
There are 1000 students (data elements), so $n = 1000$. We want to find the 56th percentile point, so $p = 56$. First, calculate the index:

$$i = (56 \times 1000)/100$$
$$= 56{,}000/100$$
$$= 560$$

This is a whole number, so we must add 0.5 to it, getting $i + 0.5 = 560.5$. This means the 56th percentile is the boundary between the "560th worst" and "561st worst" test papers. To find out what score this represents, we must check the cumulative absolute frequencies in Table 4-1. The cumulative frequency corresponding to a score of 25 is 531 (that's less than 560.5); the cumulative frequency corresponding to a score of 26 is 565 (that's more than 560.5). The 56th percentile point thus lies between scores of 25 and 26.

PROBLEM 4-2
If you happen to be among the students taking this test and you score 33, what is your percentile rank?

SOLUTION 4-2
Checking the table, you can see that 800 students have scores less than yours. (Not less than or equal to, but just less than!) This means, according to the second definition above, that $t = 800$. Then the tentative percentile $p*$ is:

$$p* = 100 \times 800/1000$$
$$= 100 \times 0.8$$
$$= 80$$

This is a whole number, so rounding it to the nearest whole number between 1 and 99 gives us $p = 80$. You are in the 80th percentile.

PROBLEM 4-3
If you happen to be among the students taking this test and you score 0, what is your percentile rank?

SOLUTION 4-3
In this case, no one has a score lower than yours. This means, according to the second definition above, that $t = 0$. The tentative percentile $p*$ is:

$$p* = 100 \times 0/1000$$
$$= 100 \times 0$$
$$= 0$$

Remember, we must round to the nearest whole number between 1 and 99 to get the actual percentile value. This is $p = 1$. Therefore, you rank in the 1st percentile.

Quartiles and Deciles

There are other ways to divide data sets besides the percentile scheme. It is common to specify points or boundaries that divide data into quarters or into tenths.

QUARTILES IN A NORMAL DISTRIBUTION

A *quartile* or *quartile point* is a number that breaks a data set up into four intervals, each interval containing approximately 1/4 or 25% of the elements in the set. There are three quartiles, not four, because the quartiles represent boundaries where four intervals meet. Thus, quartiles are assigned values 1, 2, or 3. They are sometimes called the 1st quartile, the 2nd quartile, and the 3rd quartile.

Examine Fig. 4-1 again. The location of the qth quartile point is found by locating the vertical line L such that the percentage n of the area beneath the curve is exactly equal to $25q$, and then noting the point where L crosses the horizontal axis. In Fig. 4-1, imagine that line L can be moved freely back and forth. Let n be the percentage of the area under the curve that lies to the left of L. When $n = 25\%$, line L crosses the horizontal axis at the 1st quartile point. When $n = 50\%$, line L crosses the horizontal axis at the 2nd quartile point. When $n = 75\%$, line L crosses the horizontal axis at the 3rd quartile point.

QUARTILES IN TABULAR DATA

Let's return to the 40-question test described above and in Table 4-1. Where do we put the three quartile points in this data set? There are 41 different possible scores and 1000 actual data elements. We break these 1000 results up into four different groups with three different boundary points according to the following criteria:

- The highest possible boundary point representing the "worst" 250 or fewer papers, and the 1st quartile point at the top of that set.
- The highest possible boundary point representing the "worst" 500 or fewer papers, and the 2nd quartile point at the top of that set.
- The highest possible boundary point representing the "worst" 750 or fewer papers, and the 3rd quartile point at the top of that set.

The nomograph of Fig. 4-2A illustrates the positions of the quartile points for the test results shown by Table 4-1. The data in the table are unusual; they represent a coincidence because the quartiles are all clearly defined. There are obvious boundaries between the "worst" 250 papers and the "2nd worst," between the "2nd and 3rd worst," and between the "3rd worst" and the "best." These boundaries occur at the transitions between scores of 16 and 17, 24 and 25, and 31 and 32 for the 1st, 2nd, and 3rd quartiles, respectively. If these same 1000 students are given another 40-question test, or if this 40-question test is administered to a different group of 1000 students, it's almost certain that the quartiles will not be so obvious.

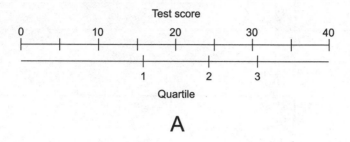

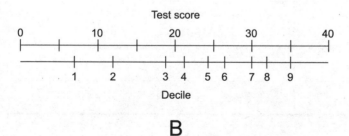

Fig. 4-2.   At A, positions of quartiles in the test results described in the text. At B, positions of the deciles.

PROBLEM 4-4
Table 4-2 shows a portion of results for the same 40-question test, but with slightly different results from those shown in Table 4-1, so that the 1st quartile point is not "cleanly" defined. Where is the 1st quartile point here?

SOLUTION 4-4
Interpret the definition literally. The 1st quartile is the *highest possible* boundary point at the top of the set of the "worst" 250 *or fewer* papers. In Table 4-2, that corresponds to the transition between scores of 16 and 17.

Table 4-2   Table for Problem 4-4.

Test score	Absolute frequency	Cumulative absolute frequency
↑	↑	↑
↑	↑	↑
↑	↑	↑
13	22	200
14	13	213
15	19	232
16	16	248
17	30	278
18	22	300
19	27	327
↓	↓	↓
↓	↓	↓
↓	↓	↓

DECILES IN A NORMAL DISTRIBUTION

A *decile* or *decile point* is a number that divides a data set into 10 intervals, each interval containing about 1/10 or 10% of the elements in the set. There are nine deciles, representing the points where the 10 sets meet. Deciles are assigned whole-number values between, and including, 1 and 9. They are sometimes called the 1st decile, the 2nd decile, the 3rd decile, and so on up to the 9th decile.

Refer again to Fig. 4-1. The location of the *d*th decile point is found by locating the vertical line L such that the percentage n of the area beneath the curve is exactly equal to $10d$, and then noting the point where L crosses the horizontal axis. Imagine again that L can be slid to the left or right at will. Let n be the percentage of the area under the curve that lies to the left of L. When $n = 10\%$, L crosses the horizontal axis at the 1st decile point. When $n = 20\%$, L crosses the horizontal axis at the 2nd decile point. When $n = 30\%$, L crosses the horizontal axis at the 3rd decile point. This continues on up, until when $n = 90\%$, line L crosses the horizontal axis at the 9th decile point.

DECILES IN TABULAR DATA

One more time, let's scrutinize the 40-question test whose results are portrayed in Table 4-1. Where do we put the decile points? We break the 1000 test papers into 10 different groups with nine different boundary points according to these criteria:

- The highest possible boundary point representing the "worst" 100 or fewer papers, and the 1st decile point at the top of that set.
- The highest possible boundary point representing the "worst" 200 or fewer papers, and the 2nd decile point at the top of that set.
- The highest possible boundary point representing the "worst" 300 or fewer papers, and the 3rd decile point at the top of that set.
  ↓
- The highest possible boundary point representing the "worst" 900 or fewer papers, and the 9th decile point at the top of that set.

The nomograph of Fig. 4-2B illustrates the positions of the decile points for the test results shown by Table 4-1. As is the case with quartiles, the data in the table are coincidental, because the deciles are obvious. There are clear boundaries between the "worst" 100 papers and the "2nd worst," between the "2nd and 3rd worst," between the "3rd and 4th worst," and so on up. If these same 1000 students are given another 40-question test, or if this 40-question test is administered to a different group of 1000 students, it's almost certain that the locations of the decile points will be less obvious. (By now you should be able to tell that this table has been contrived to make things come out neat.)

PROBLEM 4-5
Table 4-3 shows a portion of results for the same 40-question test, but with slightly different results from those portrayed in Table 4-1. Here, the 6th decile point is not "cleanly" defined. Where is that point in this case?

Table 4-3   Table for Problem 4-5.

Test score	Absolute frequency	Cumulative absolute frequency
↑	↑	↑
↑	↑	↑
↑	↑	↑
24	35	500
25	31	531
26	34	565
27	37	602
28	32	634
29	33	667
30	33	700
↓	↓	↓
↓	↓	↓
↓	↓	↓

SOLUTION 4-5

Once again, interpret the definition literally. The 6th decile is the *highest possible* boundary point at the top of the set of the "worst" 600 *or fewer* papers. In Table 4-3, that corresponds to the transition between scores of 26 and 27.

Intervals by Element Quantity

Percentiles, quartiles, and deciles can be confusing when statements are made such as, "You are in the 99th percentile of this graduating class. That's the highest possible rank." Doubtless more than one student in this elite class has asked, upon being told this, "Don't you mean to say that I'm in the 100th percentile?" After all, the term "percentile" implies there should be 100 groups, not 99.

It's all right to think in terms of the intervals between percentile boundaries, quartile boundaries, or decile boundaries, rather than the intervals centered at the boundaries. In fact, from a purely mathematical standpoint, this approach makes more sense. The 99 percentile points in a ranked data set divide that set into 100 intervals, each of which has an equal number (or as nearly an equal number as possible) of elements. Similarly, the three quartile points divide a ranked set into four intervals, as nearly equal-sized as possible; the nine decile points divide a ranked data set into 10 intervals, as nearly equal-sized as possible.

25% INTERVALS

Go back to Table 4-1 one more time, and imagine that we want to express the scores in terms of the bottom 25%, the 2nd lowest 25%, the 2nd highest 25%, and the top 25%. Table 4-4 shows the test results with the 25% intervals portrayed. They can also be called the bottom quarter, the 2nd lowest quarter, the 2nd highest quarter, and the top quarter.

Again, this particular set of scores is special because the intervals are "cleanly" defined. If things weren't so neat, we would be obliged to figure

Table 4-4   Results of a hypothetical 40-question test taken by 1000 students, with the 25% intervals indicated.

Range of scores	Absolute frequency	Cumulative absolute frequency	25% intervals
0–16	250	250	Bottom 25%
17–24	250	500	2nd Lowest 25%
25–31	250	750	2nd Highest 25%
32–40	250	1000	Top 25%

out the quartile points, and then define the 25% intervals as the sets of scores between those boundaries.

10% INTERVALS

Once again, let's revisit the test whose results are portrayed in Table 4-1. Suppose that, instead of thinking about percentiles, we want to express the scores in terms of the bottom 10%, the 2nd lowest 10%, the 3rd lowest 10%, and so forth up to the top 10%. Table 4-5 shows the test results with these intervals portrayed. The spans can also be called the 1st 10th, the 2nd 10th, the 3rd 10th, and so on up to the top 10th (or, if you want to be perverse, the 10th 10th).

Table 4-5   Results of a hypothetical 40-question test taken by 1000 students, with the 10% intervals indicated.

Range of scores	Absolute frequency	Cumulative absolute frequency	10% intervals
0–7	100	10	Bottom 10%
8–13	100	200	2nd Lowest 10%
14–18	100	300	3rd Lowest 10%
19–21	100	400	4th Lowest 10%
22–24	100	500	5th Lowest 10%
25–27	100	600	5th Highest 10%
28–30	100	700	4th Highest 10%
31–32	100	800	3rd Highest 10%
33–35	100	900	2nd Highest 10%
36–40	100	1000	Top 10%

This particular set of scores is special because the interval cutoff points are "clean." If this set was not contrived to make the discussion as easy as possible, we'd have to find the decile points, and then define the 10% intervals as the sets of scores between those boundaries.

PROBLEM 4-6

Table 4-6 shows a portion of results for a 40-question test given to 1000 students, but with slightly different results from those portrayed in Table 4-1. What range of scores represents the 2nd highest 10% in this instance?

Table 4-6   Table for Problem 4-6.

Test score	Absolute frequency	Cumulative absolute frequency
↑	↑	↑
↑	↑	↑
↑	↑	↑
30	35	702
31	51	753
32	50	803
33	40	843
34	27	870
35	31	901
36	30	930
↓	↓	↓
↓	↓	↓
↓	↓	↓

SOLUTION 4-6

The 2nd highest 10% can also be thought of as the 9th lowest 10%. It is the range of scores bounded at the bottom by the 8th decile, and at the top by the 9th decile. The 8th decile is the *highest possible* boundary point at the top of the set of the "worst" 800 *or fewer* papers. In Table 4-6, that corresponds to the transition between scores of 31 and 32. The 9th decile is the *highest possible* boundary point at the top of the set of the "worst" 900 *or fewer* papers. In Table 4-6, that corresponds to the transition between scores of 34 and 35. The 9th lowest (or 2nd highest) 10% of scores is therefore the range of scores from 32 to 34, inclusive.

PROBLEM 4-7

Table 4-7 shows a portion of results for a 40-question test given to 1000 students, but with slightly different results from those portrayed in Table 4-1. What range of scores represents the lowest 25% in this instance?

Table 4-7   Table for Problem 4-7.

Test score	Absolute frequency	Cumulative absolute frequency
↑	↑	↑
↑	↑	↑
↑	↑	↑
14	12	212
15	18	230
16	19	249
17	27	276
18	26	302
↓	↓	↓
↓	↓	↓
↓	↓	↓

SOLUTION 4-7

The lowest 25% is the range of scores bounded at the bottom by the lowest possible score, and at the top by the 1st quartile. The lowest possible score is 0. The 1st quartile is the *highest possible* boundary point at the top of the set of the "worst" 250 *or fewer* papers. In Table 4-7, that corresponds to the transition between scores of 16 and 17. The lowest 25% of scores is therefore the range of scores from 0 to 16, inclusive.

Fixed Intervals

In this chapter so far, we've divided up data sets into subsets containing equal (or as nearly equal as possible) number of elements, and then observed the ranges of values in each subset. There's another approach: we can define fixed ranges of independent-variable values, and then observe the number of elements in each range.

THE TEST REVISITED

Let's re-examine the test whose results are portrayed in Table 4-1. This time, let's think about the ranges of scores. There are many ways we can do this, three of which are shown in Tables 4-8, 4-9, and 4-10.

In Table 4-8, the results of the test are laid out according to the number of papers having scores in the following four ranges: 0–10, 11–20, 21–30, and 31–40. We can see that the largest number of students have scores in the range 21–30, followed by the ranges 31–40, 11–20, and 0–10.

Table 4-8 Results of a hypothetical 40-question test taken by 1000 students, divided into four equal ranges of scores.

Range of scores	Absolute frequency	Percentage of scores
0–10	145	14.5%
11–20	215	21.5%
21–30	340	34.0%
31–40	300	30.0%

Table 4-9 Results of a hypothetical 40-question test taken by 1000 students, divided into 10 equal ranges of scores.

Range of scores	Absolute frequency	Percentage of scores
0–4	50	5.0%
5–8	62	6.2%
9–12	66	6.6%
13–16	72	7.2%
17–20	110	11.0%
21–24	140	14.0%
25–28	134	13.4%
29–32	166	16.6%
33–36	130	13.0%
37–40	70	7.0%

Table 4-10 Results of a hypothetical 40-question test taken by 1000 students, divided into ranges of scores according to subjective grades.

Letter grade	Range of scores	Absolute frequency	Percentage of scores
F	0–18	300	30.0%
D	19–24	200	20.0%
C	25–31	250	25.0%
B	32–37	208	20.8%
A	38–40	42	4.2%

In Table 4-9, the results are shown according to the number of papers having scores in 10 ranges. In this case, the most "popular" range is 29–32. The next most "popular" range is 21–24. The least "popular" range is 0–4.

Both Tables 4-8 and 4-9 divide the test scores into equal-sized ranges (except the lowest range, which includes one extra score, the score of 0). Table 4-10 is different. Instead of breaking the scores down into ranges of equal size, the scores are tabulated according to letter grades A, B, C, D, and F. The assignment of letter grades is often subjective, and depends on the performance of the class in general, the difficulty of the test, and the disposition of the teacher. (The imaginary teacher grading this test must be a hard-nosed person.)

PIE GRAPH

The data in Tables 4-8, 4-9, and 4-10 can be portrayed readily in graphical form using broken-up circles. This is a *pie graph*, also sometimes called a *pie chart*. The circle is divided into wedge-shaped sections in the same way a pie is sliced. As the size of the data subset increases, the angular width of the pie section increases in direct proportion.

In Fig. 4-3, graph A portrays the data results from Table 4-8, graph B portrays the results from Table 4-9, and graph C portrays the results from Table 4-10. The angle at the tip or apex of each pie wedge, in degrees, is directly proportional to the percentage of data elements in the subset. Thus if a wedge portrays 10% of the students, its apex angle is 10% of $360°$, or $36°$; if a wedge portrays 25% of the students, its apex angle is 25% of $360°$, or $90°$. In general, if a wedge portrays x% of the elements in the population, the apex angle θ of its wedge in a pie graph, in degrees, is $3.6x$.

The sizes of the wedges of each pie can also be expressed in terms of the area percentage. The wedges all have the same radius – equal to the radius of the circle – so their areas are proportional to the percentages of the data elements in the subsets they portray. Thus, for example, in Fig. 4-3A, the range of scores 31–40 represents a slice containing "30% or 3/10 of the pie," while in Fig. 4-3C, we can see that the students who have grades of C represent "25% or 1/4 of the pie."

VARIABLE-WIDTH HISTOGRAM

Histograms were introduced back in Chapter 1. The example shown in that chapter is a bit of an oversimplification, because it's a *fixed-width histogram*. There exists a more flexible type of histogram, called the *variable-width*

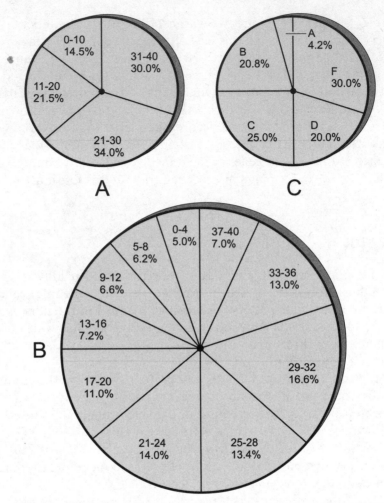

Fig. 4-3. At A, pie graph of data in Table 4-8. At B, pie graph of data in Table 4-9. At C, pie
graph of data in Table 4-10.

histogram. This sort of graph is ideal for portraying the results of our
hypothetical 40-question test given to 1000 students in various ways.

Figure 4-4 shows variable-width histograms that express the same data as
that in the tables and pie graphs. In Fig. 4-4, graph A portrays the data
results from Table 4-8, graph B portrays the results from Table 4-9, and
graph C portrays the results from Table 4-10. The width of each vertical
bar is directly proportional to the range of scores. The height of each bar
is directly proportional to the percentage of students who received scores in
the indicated range.

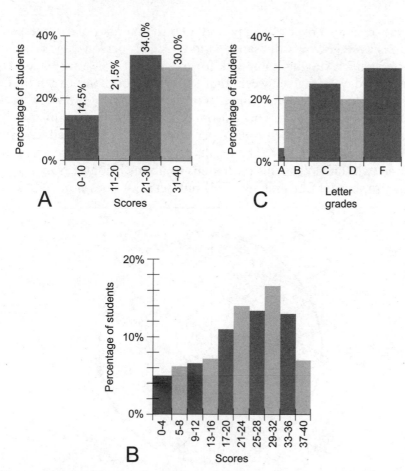

Fig. 4-4. At A, histogram of data in Table 4-8. At B, histogram of data in Table 4-9. At C, histogram of data in Table 4-10.

Percentages are included in the histogram of Fig. 4-4A, because there's room enough to show the numbers without making the graph look confusing or cluttered. In Figs. 4-4B and C, the percentages are not written at the top of each bar. This is a matter of preference. Showing the numbers in graph B would make it look too cluttered to some people. In graph C, showing the percentage for the grade of A would be difficult and could cause confusion, so they're all left out. It's a good idea to include tabular data with histograms when the percentages aren't listed at the tops of the bars.

PROBLEM 4-8
Imagine a large corporation that operates on a five-day work week (Monday through Friday). Suppose the number of workers who call in sick each day of

the week is averaged over a long period, and the number of sick-person-days per week is averaged over the same period. (A sick-person-day is the equivalent of one person staying home sick for one day. If the same person calls in sick for three days in a given week, that's three sick-person-days in that week, but it's only one sick person.) For each of the five days of the work week, the average number of people who call in sick on that day is divided by the average number of sick-person-days per week, and is tabulated as a percentage for that work-week day. The results are portrayed as a pie graph in Fig. 4-5. Name two things that this graph tells us about Fridays. Name one thing that this graph might at first seem to, but actually does not, tells us about Fridays.

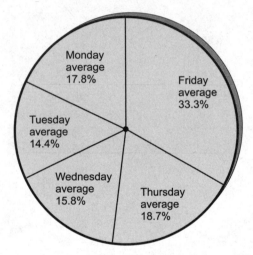

Fig. 4-5.   Illustration for Problems 4-8 through 4-10.

SOLUTION 4-8
The pie graph indicates that more people (on the average) call in sick on Fridays than on any other day of the work week. It also tells us that, of the total number of sick-person-days on a weekly basis, an average of 33% of them occur on Fridays. The pie graph might at first seem to, but in fact does not, indicate that an average of 33% of the workers in the corporation call in sick on Fridays.

PROBLEM 4-9
Suppose that, in the above described corporation and over the survey period portrayed by the pie graph of Fig. 4-5, there are 1000 sick-person-days per week on average. What is the average number of sick-person-days on

Mondays? What is the average number of people who call in sick on Mondays?

SOLUTION 4-9
For a single day, a sick-person-day is the equivalent of one person calling in sick. But this is not necessarily true for any period longer than one day. In this single-day example, we can multiply 1000 by 17.8%, getting 178, and this gives us both answers. There are, on the average, 178 sick-person-days on Mondays. An average of 178 individuals call in sick on Mondays.

PROBLEM 4-10
Given the same scenario as that described in the previous two problems, what is the average number of sick-person-days on Mondays and Tuesdays combined? What is the average number of individuals who call in sick on Mondays and Tuesdays combined?

SOLUTION 4-10
An average of 178 sick-person-days occur on Mondays, as we have determined in the solution to the previous problem. To find the average number of sick-person-days on Tuesdays, multiply 1000 by 14.4%, getting 144. The average number of sick-person-days on Mondays and Tuesdays combined is therefore 178 + 144, or 322. It is impossible to determine the average number of individuals who call in sick on Mondays and Tuesdays combined, because we don't know how many of the Monday–Tuesday sick-person-day pairs represent a single individual staying out sick on both days (two sick-person-days but only one sick person).

Other Specifications

There are additional descriptive measures that can be used to describe the characteristics of data. Let's look at the definitions of some of them.

RANGE

In a data set, or in any contiguous ("all-of-a-piece") interval in that set, the term *range* can be defined as the difference between the smallest value and the largest value in the set or interval.

In the graph of hypothetical blood-pressure test results (Fig. 4-1), the lowest systolic pressure in the data set is 60, and the highest is 160.

Therefore, the range is the difference between these two values, or 100. It's possible that a few of the people tested have pressures lower than 60 or higher than 160, but their readings have been, in effect, thrown out of the data set.

In the 40-question test we've examined so often in this chapter, the lowest score is 0, and the highest score is 40. Therefore, the range is 40. We might want to restrict our attention to the range of some portion of all the scores, for example the 2nd lowest 25% of them. This range can be determined from Table 4-4; it is equal to $24 - 17$, or 7. Note that the meaning of the word "range" in this context is different from the meaning of the word "range" at the top of the left-hand column of Table 4-4.

COEFFICIENT OF VARIATION

Do you remember the definitions of the mean (μ) and the standard deviation (σ) from Chapter 2? Let's review them briefly. There's an important specification that can be derived from them.

In a normal distribution, such as the one that shows the results of our hypothetical blood-pressure data-gathering adventure, the mean is the value (in this case the blood pressure) such that the area under the curve is equal on either side of a vertical line corresponding to that value.

In tabulated data for discrete elements, the mean is the arithmetic average of all the results. If we have results $\{x_1, x_2, x_3, \ldots, x_n\}$ whose mean is μ, then the standard deviation is

$$\sigma = \{(1/n)[(x_1 - \mu)^2 + (x_2 - \mu)^2 + \ldots + (x_n - \mu)^2]\}^{1/2}$$

The mean is a measure of central tendency or "centeredness." The standard deviation is a measure of dispersion or "spread-outedness." Suppose we want to know how spread out the data is relative to the mean. We can get an expression for this if we divide the standard deviation by the mean. This gives us a quantity known as the *coefficient of variation*, which is symbolized as CV. Mathematically, the CV is:

$$CV = \sigma/\mu$$

The standard deviation and the mean are both expressed in the same units, such as systolic blood pressure or test score. Because of this, when we divide one by the other, the units cancel each other out, so the CV doesn't have any units. A number with no units associated with it is known as a *dimensionless quantity*.

Because the CV is dimensionless, it can be used to compare the "spread-outedness" of data sets that describe vastly different things, such as blood pressures and test scores. A large CV means that data is relatively spread out around the mean. A small CV means that data is concentrated closely around the mean. In the extreme, if CV = 0, all the data values are the same, and are exactly at the mean. Figure 4-6 shows two distributions in graphical form, one with a fairly low CV, and the other with a higher CV.

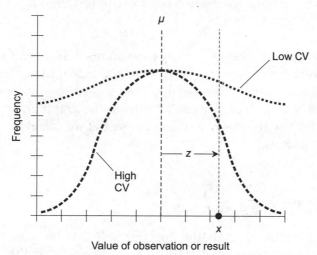

Fig. 4-6. Two distributions shown in graphical form, one with a low coefficient of variation (CV) and one with a higher CV. Derivation of the Z score (z) for result x is discussed in the text.

There is one potential difficulty with the above formula. Have you guessed it? If you wonder what happens in a distribution where the data can attain either positive or negative values – for example, temperatures in degrees Celsius – your concern is justified. If $\mu = 0$ (the freezing point of water on the Celsius temperature scale), there's a problem. This trouble can be avoided by changing the units in which the data is specified, so that 0 doesn't occur within the set of possible values. When expressing temperatures, for example, we could use the Kelvin scale rather than the Celsius scale, where all temperature readings are above 0.

In a situation where all the elements in a data set are equal to 0, such as would happen if a whole class of students turns in blank papers on a test, the CV is undefined because the mean really is equal to 0.

Z SCORE

Sometimes you'll hear people say that such-and-such an observation or result is "2.2 standard deviations below the mean" or "1.6 standard deviations above the mean." The Z *score*, symbolized z, is a quantitative measure of the position of a particular element with respect to the mean. The Z score of an element is equal to the number of standard deviations that the element differs from the mean, either positively or negatively.

For a specific element x in a data set, the value of z depends on both the mean (μ) and the standard deviation (σ) and can be found from this formula:

$$z = (x - \mu)/\sigma$$

If x is below the mean, then z is a negative number. If x is above the mean, then z is a positive number. If x is equal to the mean, then $z = 0$.

In the graphical distributions of Fig. 4-6, $z > 0$ for the point x shown. This is true for both curves. We can't tell from the graph alone exactly what the Z score is for x with respect to either curve, but we can at least see that it's positive in both cases.

INTERQUARTILE RANGE

Sometimes it's useful to know the "central half" of the data in a set. The *interquartile range*, abbreviated IQR, is an expression of this. The IQR is equal to the value of the 3rd quartile point minus the value of the 1st quartile point. If a quartile point occurs between two integers, it can be considered as the average of the two integers (the smaller one plus 0.5).

Consider again the hypothetical 40-question test taken by 1000 students. The quartile points are shown in Fig. 4-2A. The 1st quartile occurs between scores of 16 and 17; the 3rd quartile occurs between scores of 31 and 32. Therefore:

$$IQR = 31 - 16$$
$$= 15$$

PROBLEM 4-11
Suppose a different 40-question test is given to 1000 students, and the results are much more closely concentrated than those from the test depicted in Fig. 4-2A. How would the IQR of this test compare with the IQR of the previous test?

SOLUTION 4-11
The IQR would be smaller, because the 1st and 3rd quartiles would be closer together.

PROBLEM 4-12

Recall the empirical rule from the previous chapter. It states that all normal distributions have the following three characteristics:

- Approximately 68% of the data points are within the range $\pm\sigma$ of μ.
- Approximately 95% of the data points are within the range $\pm2\sigma$ of μ.
- Approximately 99.7% of the data points are within the range $\pm3\sigma$ of μ.

Re-state this principle in terms of Z scores.

SOLUTION 4-12

As defined above, the Z score of an element is the number of standard deviations that the element departs from the mean, either positively or negatively. All normal distributions have the following three characteristics:

- Approximately 68% of the data points have Z scores between -1 and $+1$.
- Approximately 95% of the data points have Z scores between -2 and $+2$.
- Approximately 99.7% of the data points have Z scores between -3 and $+3$.

Quiz

Refer to the text in this chapter if necessary. A good score is 8 correct. Answers are in the back of the book.

1. Suppose a large number of people take a test. The 3rd decile point is determined by
   (a) finding the highest score representing the "worst" 20% or fewer papers; the 3rd decile is at the top of that set
   (b) finding the highest score representing the "worst" 30% or fewer papers; the 3rd decile is at the top of that set
   (c) finding the lowest score representing the "best" 20% or fewer papers; the 3rd decile is at the bottom of that set
   (d) finding the lowest score representing the "best" 30% or fewer papers; the 3rd decile is at the bottom of that set

2. Suppose many students take a 10-question quiz, and the mean turns out to be 7.22 answers correct. Suppose the standard deviation turns out to be 0.722. What is the coefficient of variation?

(a) We can't answer this unless we know how many people take the quiz.

(b) 0.1

(c) 10

(d) 100

3. Suppose several students take a 10-question quiz. The worst score is 3 correct, and the best score is 10 correct. What is the range?

(a) We can't answer this unless we know how many people take the quiz.

(b) 3/7

(c) 7

(d) 7/3

4. Suppose several students take a 10-question quiz. The worst score is 3 correct, and the best score is 10 correct. What is the 50th percentile?

(a) 7, which is equal to $10 - 3$.

(b) 6.5, which is equal to $(3 + 10)/2$.

(c) $30^{1/2}$, which is equal to the square root of (3×10).

(d) We can't answer this unless we know how many students receive each score.

5. Imagine that you take a standardized test, and after you've finished you are told that you are at the 91st decile. This means

(a) 90 of the students taking the same test have scores higher than yours

(b) 90% of all the students taking the same test have scores higher than yours

(c) 90 of the students taking the same test have scores lower than yours

(d) nothing; there is no such thing as the 91st decile

6. Table 4-11 shows the results of a hypothetical 10-question test given to a group of students. Where is the 1st quartile point?

(a) At the transition between scores of 1 and 2.

(b) At the transition between scores of 4 and 5.

(c) At the transition between scores of 6 and 7.

(d) It cannot be determined without more information.

7. What is the range for the scores achieved by students in the scenario of Table 4-11?

(a) 5

(b) 8

Table 4-11   Illustration for Quiz Questions 6 through 9. Results of a hypothetical 10-question test taken by 128 students.

Test score	Absolute frequency	Cumulative absolute frequency
0	0	0
1	0	0
2	3	3
3	8	11
4	12	23
5	15	38
6	20	58
7	24	82
8	21	103
9	14	117
10	11	128

    (c) 10
    (d) It cannot be determined without more information.

8. What is the interquartile range of the scores in Table 4-11?
    (a) 3
    (b) 4
    (c) 6
    (d) It cannot be determined because it is ambiguous.

9. Suppose, in the scenario shown by Table 4-11, Jim Q. is one of the students taking the test, and he gets a score of 6. In what interval is Jim Q. with respect to the class?
    (a) The bottom 25%.
    (b) The next-to-lowest 25%.

(c) The next-to-highest 25%.

(d) It cannot be determined because it is ambiguous.

10. Suppose you see a pie graph showing the results of a survey. The purpose of the survey is to determine the number and proportion of families in a certain city earning incomes in various ranges. Actual numbers are, for some reason, not indicated on this pie graph. But you see that one of the ranges has a "slice" with an apex (central) angle of 90°. From this, you can assume that the slice corresponds to

(a) the families whose earnings fall into the middle 25% of the income range

(b) 25% of the families surveyed

(c) $1/\pi$ of the families surveyed (π is the ratio of a circle's circumference to its diameter)

(d) the families whose earnings fall within the interquartile range

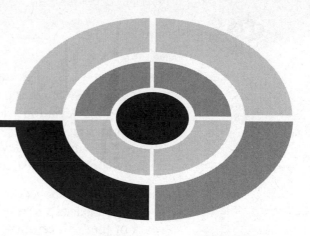

Test: Part One

Do not refer to the text when taking this test. You may draw diagrams or use a calculator if necessary. A good score is at least 45 correct. Answers are in the back of the book. It's best to have a friend check your score the first time, so you won't memorize the answers if you want to take the test again.

1. If you take a standardized test and then you are told you are in the 50th percentile, this means
    (a) that your score is among the lowest in the range
    (b) that your score is among the best in the range
    (c) that your score is near the middle of the range
    (d) that your score has low correlation
    (e) nothing in particular about your score

2. How many decile points are there in a set of 100,000 ranked data elements?
    (a) More information is necessary in order to answer this question.
    (b) 9
    (c) 10
    (d) 99
    (e) 100

3. The term *fuzzy truth* is used to describe
   (a) a theory in which there are degrees of truth that span a range
   (b) standard deviation
   (c) cumulative frequency
   (d) the probability fallacy
   (e) any continuous distribution

4. Figure Test 1-1 shows the results of sunshine research in five imaginary towns, based on research carried out daily over the entire 100 years of the 20th century. What type of data portrayal is this?
   (a) A horizontal bar graph.
   (b) A point-to-point graph.
   (c) A correlation chart.
   (d) A cumulative frequency graph.
   (e) A pie chart.

5. What, if anything, is mathematically wrong or suspect in Fig. Test 1-1?
   (a) The bars should get longer and longer as you go further down.
   (b) All the bars should be the same length.
   (c) All the numbers at the right-hand ends of the bars should add up to 100%.
   (d) All the numbers at the right-hand ends of the bars should add up to the average number of days in a year (approximately 365.25).
   (e) There is nothing mathematically wrong or suspect in Fig. Test 1-1.

6. If the numbers at the right-hand ends of the bars in Fig. Test 1-1 represent percentages of days in an average year over the course of

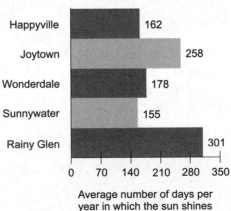

Fig. Test 1-1.   Illustration for Part One Test Questions 4 through 6.

the observation period (rather than the actual number of days in an average year), how will the numbers be different? Assume the average number of days in a year is 365.25.

(a) All the numbers will be reduced by a factor of 3.6525, and will be followed by percent symbols.

(b) The sum of all the numbers will have to equal 100%.

(c) All the numbers will be increased by a factor of 3.6525, and will be followed by percent symbols.

(d) All the numbers will remain unchanged, but will be followed by percent symbols.

(e) Fig. Test 1-1 does not contain enough information to determine percentages of days in an average year.

7. The average of the outcomes in an experiment is known as the
   (a) continuous variable
   (b) discrete variable
   (c) random variable
   (d) mean
   (e) frequency

8. Which of the following pairs of characteristics are both measures of the extent to which the data in a distribution is spread out?
   (a) The mean and the median.
   (b) The mean and the deviation.
   (c) The variance and the standard deviation.
   (d) The mode and the mean.
   (e) None of the above.

9. What is the mathematical probability that a coin, tossed 13 times in a row, will come up "heads" on all 13 tosses?
   (a) 1 in 512
   (b) 1 in 1024
   (c) 1 in 2048
   (d) 1 in 4096
   (e) 1 in 8192

10. A member of a set is also called
    (a) a dependent variable of the set
    (b) an independent variable of the set
    (c) a random variable of the set
    (d) a discrete variable of the set
    (e) none of the above

11. A continuous variable can attain
    (a) no values
    (b) one value
    (c) a positive whole number of values
    (d) infinitely many values
    (e) only negative values

12. Fill in the blank in the following sentence to make it true: "In a normal distribution, a _____ is a number that divides a data set into 10 intervals, each interval containing about 1/10 or 10% of the elements in the data set."
    (a) range
    (b) coefficient
    (c) decile
    (d) median
    (e) mode

13. When an object x is in either set A or set B but not both, then we can be sure that
    (a) $x \in A \cup B$
    (b) $x \in A \cap B$
    (c) $x \geq A$
    (d) $x \notin B$
    (e) $B(x) \in \varnothing$

14. Suppose a large number of students take this test, and the results are portrayed in Fig. Test 1-2. This is an example of
    (a) a point-to-point graph
    (b) a continuous-curve graph
    (c) a histogram
    (d) a horizontal-bar graph
    (e) a normal distribution

15. In a graph of the type shown in Fig. Test 1-2, it is important that
    (a) the values all add up to 100%
    (b) no single value exceeds 50%
    (c) no two values be the same
    (d) the independent variable be shown on the vertical axis
    (e) numbers always be shown at the tops of the bars

16. In a graph of the type shown in Fig. Test 1-2, what is the maximum possible height that a bar could have (that is, the largest possible percentage)?

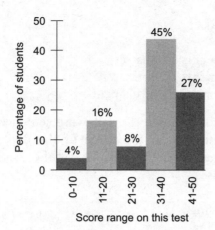

Fig. Test 1-2. Illustration for Part One Test Questions 14 through 16.

    (a) 0%
    (b) 20%
    (c) 50%
    (d) 80%
    (e) 100%

17. The set of integers has elements that are all
    (a) positive whole numbers or negative whole numbers
    (b) positive whole numbers, negative whole numbers, or zero
    (c) quotients of positive whole numbers
    (d) quotients of positive whole numbers or zero
    (e) quotients of negative whole numbers

18. Mathematical probabilities can have values
    (a) between -1 and 1 inclusive
    (b) corresponding to any positive real number
    (c) between 0 and 1 inclusive
    (d) corresponding to any integer
    (e) within any defined range

19. An outcome is the result of
    (a) a discrete variable
    (b) an independent variable
    (c) a correlation
    (d) a population
    (e) an event

20. The intersection of two disjoint sets contains

(a) no elements
(b) one element
(c) all the elements of one set
(d) all the elements of both sets
(e) an infinite number of elements

21. Fill in the blank to make the following sentence true: "A function is
_____ if and only if the value of the dependent variable never
grows any larger (or more positive) as the value of the independent
variable increases."
(a) random
(b) nondecreasing
(c) nonincreasing
(d) constant
(e) trending horizontally

22. Fill in the blank to make the following sentence true: "If the number
of elements in a distribution is even, then the _____ is the value
such that half the elements have values greater than or equal to it, and
half the elements have values less than or equal to it."
(a) mean
(b) average
(c) standard deviation
(d) median
(e) variance

23. Figure Test 1-3 shows a general illustration of
(a) a normal distribution
(b) an invariant distribution
(c) a random distribution
(d) a uniform distribution
(e) a variant distribution

24. In Fig. Test 1-3, the symbol σ represents
(a) the distribution
(b) the variance
(c) the mean
(d) the median
(e) none of the above

25. The curve shown in Fig. Test 1-3 is often called
(a) a linear function
(b) a quadratic curve

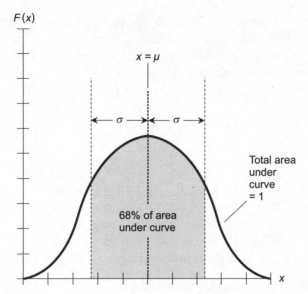

Fig. Test 1-3. Illustration for Part One Test Questions 23 through 25.

    (c) a parabola
    (d) a bell-shaped curve
    (e) a complex curve

26. Let q represent a set of items or objects taken r at a time in no particular order, where both q and r are positive integers. The possible number of combinations in this situation is symbolized $_qC_r$ and can be calculated as follows:

$$_qC_r = q! \ / \ [r!(q - r)!]$$

Given this information, what is the possible number of combinations of 150 objects taken 2 at a time?
    (a) It is a huge number, and cannot be calculated precisely in a reasonable length of time without a computer.
    (b) 22,350
    (c) 11,175
    (d) 150
    (e) 148

27. Let q represent a set of items or objects taken r at a time in a specific order. The possible number of permutations in this situation is symbolized $_qP_r$ and can be calculated as follows:

$$_qP_r = q! \ / \ (q - r)!$$

Given this information, what is the possible number of permutations of 150 objects taken 2 at a time?

(a) It is a huge number, and cannot be calculated precisely in a reasonable length of time without a computer.

(b) 22,350

(c) 11,175

(d) 150

(e) 148

28. Which of the following is an example of a discrete variable?

(a) The direction of the wind as a tornado passes.

(b) The number of car accidents per month in a certain town.

(c) The overall loudness of sound during a symphony.

(d) The speed of a car on a highway.

(e) The thrust of a jet engine during an airline flight.

29. Two outcomes are independent if and only if

(a) they always occur simultaneously

(b) one occurs only when the other does not

(c) they sometimes occur simultaneously, but usually they do not

(d) the occurrence of one does not affect the probability that the other will occur

(e) they rarely occur simultaneously, but once in a while they do

30. How many 25% intervals are there in a set of 100 ranked data elements?

(a) 3

(b) 4

(c) 9

(d) 10

(e) 99

31. When manipulating an equation, which of the following actions is not allowed?

(a) Multiplication of both sides by the same constant.

(b) Subtraction of the same constant from both sides.

(c) Addition of the same constant to both sides.

(d) Division of both sides by a variable that may attain a value of zero.

(e) Addition of a variable that may attain a value of zero to both sides.

32. In a frequency distribution:

(a) the frequency is always less than 0
(b) there is only one possible frequency
(c) frequency is portrayed as the independent variable
(d) frequency is portrayed as the dependent variable
(e) none of the above

33. A graph that shows proportions that look like slices of a pizza is called
    (a) a histogram
    (b) a slice graph
    (c) a pie graph
    (d) a bar graph
    (e) a nomograph

34. In Fig. Test 1-4, suppose H_1 and H_2 represent two different sets of outcomes in an experiment. The light-shaded region, labeled $H_1 \cap H_2$, represents
    (a) the set of outcomes common to both H_1 and H_2
    (b) the set of outcomes belonging to neither H_1 nor H_2
    (c) the set of outcomes belonging to either H_1 or H_2, but not both
    (d) the set of outcomes belonging to either H_1 or H_2, or both
    (e) the empty set

35. In Fig. Test 1-4, the dark-shaded region, labeled $H_1 \cup H_2 - H_1 \cap H_2$, shows the set of elements that are in $H_1 \cup H_2$ but not in $H_1 \cap H_2$. In a statistical experiment, this can represent
    (a) the set of outcomes common to both H_1 and H_2
    (b) the set of outcomes belonging to neither H_1 nor H_2
    (c) the set of outcomes belonging to either H_1 or H_2, but not both
    (d) the set of outcomes belonging to either H_1 or H_2, or both
    (e) the empty set

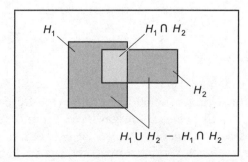

Fig. Test 1-4. Illustration for Part One Test Questions 34 through 37.

36. In Fig. Test 1-4, the entire portion that is shaded, either light or dark, represents
    (a) the set of outcomes common to both H_1 and H_2
    (b) the set of outcomes belonging to neither H_1 nor H_2
    (c) the set of outcomes belonging to either H_1 or H_2, but not both
    (d) the set of outcomes belonging to either H_1 or H_2, or both
    (e) the empty set

37. The Venn diagram of Fig. Test 1-4 portrays
    (a) complementary outcomes
    (b) mutually exclusive outcomes
    (c) independent outcomes
    (d) coincident outcomes
    (e) nondisjoint outcomes

38. In a normal distribution, an element in the 35th percentile lies within
    (a) the 1st quartile
    (b) the 2nd quartile
    (c) the 3rd quartile
    (d) the 4th quartile
    (e) the middle quartile

39. For any given positive integer n, the value of $(n + 1)!$ is always
    (a) larger than $n!$
    (b) smaller than $n!$
    (c) equal to $n!$
    (d) a whole-number fraction of $n!$
    (e) none of the above

40. What is the mathematical probability that an "unweighted" die, tossed four times, will show the face with 6 dots on all four occasions?
    (a) 1 in 6
    (b) 1 in 36
    (c) 1 in 64
    (d) 1 in 1296
    (e) 1 in 46,656

41. Tables Test 1-1 and Test 1-2 portray the results of a hypothetical experiment consisting of 6000 tosses of five different dice. In each of the 6000 events, all five dice are gathered up and thrown at the same time. What is a fundamental difference between these two tables?
    (a) Table Test 1-1 shows ungrouped data, and Table Test 1-2 shows grouped data.

Table Test 1-1 Illustration for Part One
Test Questions 41 through 43.

Face of die	Toss results for all dice
1	4967
2	5035
3	4973
4	4989
5	5007
6	5029

Table Test 1-2 Illustration for Part One Test Questions 41, 44, and 45.

Face of die	Toss results by manufacturer				
	Red Corp. A	Orange Corp. B	Yellow Corp. C	Green Corp. D	Blue Corp. E
1	1000	967	1000	1000	1000
2	1000	1035	1000	1000	1000
3	1000	973	1000	1000	1000
4	1000	989	1000	1000	1000
5	1000	1007	1000	1000	1000
6	1000	1029	1000	1000	1000

(b) Table Test 1-1 shows grouped data, and Table Test 1-2 shows ungrouped data.

(c) Table Test 1-1 shows weighted data, and Table Test 1-2 shows unweighted data.

(d) Table Test 1-1 shows unweighted data, and Table Test 1-2 shows weighted data.

(e) There is no functional difference between the two tables.

42. What general conclusion can be drawn from Table Test 1-1?

(a) One of the five dice is heavily "weighted," but the other four are not.

(b) Three of the five dice are heavily "weighted," and the other three are not.

(c) The group of five dice, taken together, appears to be heavily "weighted" with a strong bias toward the higher face numbers.

(d) The group of five dice, taken together, appears to be essentially "unweighted" with no significant bias toward any of the face numbers.

(e) The table has a mistake because the numbers don't add up right.

43. Suppose the experiment whose results are portrayed in Table Test 1-1 is repeated, and another table of the same format is compiled. What should we expect?

(a) Each die face should turn up approximately 5000 times.

(b) Some of the die faces should turn up far more than 5000 times, while others should turn up far less than 5000 times.

(c) The exact same results as those shown in Table Test 1-1 should be obtained.

(d) The faces that turned up less than 5000 times in the first experiment should have a tendency to turn up more than 5000 times in the second experiment, and the faces that turned up more than 5000 times in the first experiment should have a tendency to turn up less than 5000 times in the second experiment.

(e) We can't say anything about what to expect.

44. Table Test 1-2 shows the results of the same experiment as is shown by Fig. Test 1-1, but the data is more detailed. The dice are named by color (red, orange, yellow, green, blue) and by manufacturer (Corporations A, B, C, D, and E). What can be said about this table?

(a) The numbers don't add up right.

(b) A coincidence like this cannot possibly occur.

(c) The orange die is heavily "weighted" and the other five are "unweighted."

(d) The orange die is "unweighted" and the other five are heavily "weighted."

(e) It represents a perfectly plausible scenario.

45. Suppose the experiment whose results are portrayed in Table Test 1-2 is repeated, and another table of the same format is compiled. What should we expect?

(a) One die (but not the orange one) should show some variability, but all the other dice should show results of 1000 for each of their faces.

(b) Some of the die faces should turn up far more than 1000 times, while others should turn up far less than 1000 times.

(c) Each face of every die should turn up exactly 1000 times.

(d) Each face of every die should turn up approximately 1000 times.

(e) We can't say anything about what to expect.

46. A variable-width histogram is an excellent scheme for showing

(a) proportions

(b) correlation

(c) medians

(d) variances

(e) ranges

47. When the values of a function are shown on a coordinate system for selected points, and adjacent pairs of points are connected by straight lines, the resulting illustration is

(a) a quadratic graph

(b) a bar graph

(c) a Venn diagram

(d) a histogram

(e) a point-to-point graph

48. Examine Fig. Test 1-5. The points represent actual temperature readings, in degrees Celsius (°C), taken at 6-hour intervals over the course of a hypothetical day. The heavy dashed line is an educated guess of the actual function of temperature versus time during that day. This guess is an example of

(a) functional extrapolation

(b) curve fitting

(c) variable extrapolation

(d) linear interpolation

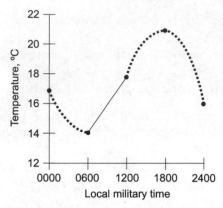

Fig. Test 1-5.   Illustration for Part One Test Questions 48 through 50.

(e) point shifting

49. In Fig. Test 1-5, suppose the heavy dashed line represents actual temperature readings obtained at 5-minute intervals during the course of a day, except for a 6-hour gap between 0600 and 1200. The straight line represents a crude attempt to fill in this gap, and is known as
(a) functional extrapolation
(b) curve fitting
(c) variable extrapolation
(d) linear interpolation
(e) point shifting

50. In Fig. Test 1-5, time represents the
(a) independent variable
(b) dependent variable
(c) curve variable
(d) continuous random variable
(e) discrete random variable

51. In a ranked data set, the value of the 3rd quartile point minus the value of the 1st quartile point is called
(a) the interquartile range
(b) the standard deviation
(c) the variance
(d) the coefficient of variation
(e) the Z score

52. If two outcomes H_1 and H_2 are complementary, then the probability, expressed as a ratio, of one outcome is equal to

    (a) the probability of the other outcome
    (b) 1 times the probability of the other outcome
    (c) 1 divided by the probability of the other outcome
    (d) 1 plus the probability of the other outcome
    (e) 1 minus the probability of the other outcome

53. A sample of a population is
    (a) an experiment in the population
    (b) a subset of the population
    (c) a variable in the population
    (d) an outcome of the population
    (e) a correlation within the population

54. The collection of data with the intent of discovering something is called
    (a) a population
    (b) an experiment
    (c) a random variable
    (d) a discrete variable
    (e) a continuous variable

55. Suppose set C is a proper subset of set D. The union of these two sets
    (a) is the same as set C
    (b) contains no elements
    (c) contains one element
    (d) is the same as set D
    (e) contains infinitely many elements

56. Suppose you are among a huge group of students who have taken a standardized test. You are told that you scored better than 97 out of every 100 students. From this, you can gather that you are in the
    (a) 99th decile
    (b) 97th decile
    (c) 10th decile
    (d) 9th decile
    (e) 2nd quartile

57. When two variables are strongly correlated and their values are plotted as points on a graph, the points
    (a) lie along a well-defined line
    (b) are clustered near the top of the graph
    (c) are clustered near the bottom of the graph
    (d) are clustered near the center of the graph

(e) are scattered all over the graph

58. If the fact that x is an element of set Q implies that x is also an element of set P, then
(a) P is a subset of Q
(b) P is a member of Q
(c) Q is a subset of P
(d) Q is a member of P
(e) none of the above

59. Fill in the blank to make the following sentence true: "Percentile points represent the 99 boundaries where _____ intervals meet."
(a) 25%
(b) 10%
(c) 101
(d) 100
(e) 99

60. When a function has a graph in which time is the independent variable, short-term forecasting can sometimes be done by
(a) interpolation
(b) Venn diagramming
(c) histography
(d) functional inversion
(e) extrapolation

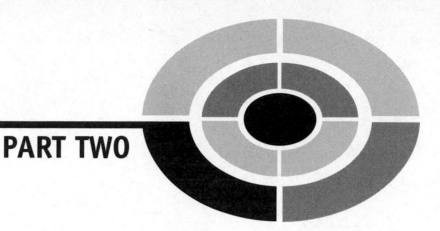

PART TWO

Statistics in Action

CHAPTER

Sampling and Estimation

In the real world, data is gathered by a process called *sampling*. It's important that the sampling process be carried out correctly, and that errors of all kinds be minimized (unless the intent is to deceive).

Source Data and Sampling Frames

When conducting a statistical experiment, four steps are followed. They must be done in order. Here they are:

1. Formulate the question. What do we want to know, and what (or who) do we want to know it about?
2. Gather the data from the required places, from the right people, and over a sufficient period of time.

3. Organize and analyze the data, so it becomes *information*.
4. Interpret the information gathered and organized from the experiment, so it becomes *knowledge*.

PRIMARY VERSUS SECONDARY

If no data are available for analysis, we have to collect it ourselves. Data collected by the statistician is called *primary source data*. If data are already available and all a statistician has to do is organize it and analyze it, then it is called *secondary source data*. There are certain precautions that must be taken when using either kind of data.

In the case of primary source data, we must be careful to follow the proper collection schemes, and then we have to be sure we use the proper methods to organize, evaluate, and interpret it. That is, we have to ensure that each of the above Steps 2, 3, and 4 are done properly. With secondary source data, the collection process has already been done for us, but we still have to organize, evaluate, and interpret it, so we have to carry out Steps 3 and 4. Either way, there's plenty to be concerned about. There are many ways for things to go wrong with an experiment, but only one way to get it right.

SAMPLING FRAMES

The most common data-collection schemes involve obtaining samples that represent a population with minimum (and ideally no) bias. This is easy when the population is small, because then the entire population can be sampled. However, a good sampling scheme can be difficult to organize when a population is large, and especially when it is not only huge but is spread out over a large region or over a long period of time.

In Chapter 2, we learned that the term *population* refers to a particular set of items, objects, phenomena, or people being analyzed. An example of a population is the set of all the insects in the world. The term *sample* was also defined in Chapter 2. A sample of a population is a subset of that population. Consider, as a sample from the foregoing population, the set of all the mosquitoes in the world that carry malaria.

It can be useful in some situations to define a set that is intermediate between a sample and a population. This is often the case when a population is huge. A *sampling frame* is a set of items within a population from which a sample is chosen. The idea is to whittle down the size of the sample, while still obtaining a sample that fairly represents the population. In the mosquito experiment, the sampling frame might be the set of all mosquitoes caught by a team of researchers, one for each 10,000 square kilometers (10^4 km^2) of land surface area in the world, on the first day of each month for one complete calendar year. We could then test all the recovered insects for the presence of the malaria protozoan.

In the simplest case, the sampling frame coincides with the population (Fig. 5-1A). However, in the mosquito experiment described above, the sampling frame is small in comparison with the population (Fig. 5-1B). Occasionally, a population is so large, diverse, and complicated that two sampling frames might be used, one inside the other (Fig. 5-1C). If the number of mosquitoes caught in the above process is so large that it would take too much time to individually test them all, we could select, say, 1% of the mosquitoes at random from the ones caught, and test each one of them.

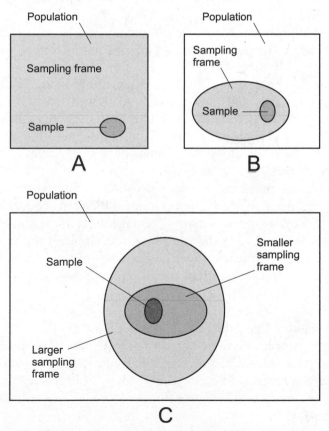

Fig. 5-1.   At A, the sampling frame coincides with the population. At B, the sampling frame is a proper subset of the population. At C, there are two sampling frames, one inside the other.

CHOOSING FRAMES

The choice of sampling frames is important, because each frame must be a fair (unbiased) representation of the next larger element set in which it is contained. Let's look at another frames-within-frames example that can be descriptive, even though it does not lend itself to illustration.

Imagine that we want to evaluate some characteristic of real numbers. The population is the set of all real numbers. Sampling frames are a matter of choice. How about the irrational numbers? Or the rational numbers? How about the set of all real numbers that are square roots of whole numbers? Suppose we choose the set of rational numbers as the sampling frame. Within this set, we might further specify subframes. How about the set of integers? Or the set of rational numbers whose quotients have denominators that are natural numbers between and including 1 and 100? How about the set of even integers? Or do you prefer the set of odd integers? Finally, within this set, we choose a sample. How about the set of integers divisible by 100? Or the set of odd integers that are 1 greater than every integer divisible by 100?

Throughout this process, we must keep one thing in mind: All the sampling frames we choose, and the final sample as well, must be an unbiased representation of the population for the purposes of our experiment. Depending on what this purpose happens to be, the whittling-down process we choose might be satisfactory, or it might be questionable, or it might put us way off the track.

In any real-life experiment, the sample should not be too large or too small. If the sample is too large, it becomes difficult to collect all the data because the process takes too many human-hours, or requires too much travel, or costs too much. If the sample is too small, it will not be a fair representation of the population for the purposes of the experiment. As the sample gets smaller, the risk of its being a poor representation increases.

PROBLEM 5-1

Suppose you want to describe the concept of a "number" to pre-college school students. In the process of narrowing down sets of numbers described above into sampling frames in an attempt to make the idea of a number clear to a child, name a few possible assets, and a few limitations.

SOLUTION 5-1

Think back to when you were in first grade. You knew what a whole number is. The concept of whole number might make a good sampling frame when talking about the characteristics of a number to a six-year-old. But by the time you were in third grade, you knew about fractions, and therefore about rational numbers. So the set of whole numbers would not have been a large enough sampling frame to satisfy you at age eight. But try talking about irrational numbers to a third grader! You won't get far! A 12th-grader would (we hope) know all about the real numbers and various subcategories of numbers within it. Restricting the sampling frame to the rational numbers would leave a 12th-grader unsatisfied. Beyond the real numbers are the realms of the complex numbers, vectors, quaternions, tensors, and transfinite numbers.

PROBLEM 5-2
Suppose you want to figure out the quantitative effect (if any) that cigarette smoking has on people's blood pressure. You conduct the experiment on a worldwide basis, for all races of people, female and male. You interview people and ask them how much they smoke, and you measure their blood pressures. The population for your experiment is the set of all people in the world. Obviously you can't carry out this experiment for this entire population! Suppose you interview 100 people from each country in the world. The resulting group of people constitutes the sampling frame. What are some of the possible flaws with this scheme? Pose the issues as questions.

SOLUTION 5-2
Here are some questions that would have to be answered, and the issues resolved, before you could have confidence in the accuracy of this experiment.

- How do you account for the fact that some countries have far more people than others?
- How do you account for the fact that the genetic profiles of the people in various countries differ?
- How do you account for the fact that people smoke more in some countries than in others?
- How do you account for the fact that the average age of the people in various countries differs, and age is known to affect blood pressure?
- How do you account for differences in nutrition in various countries, a factor that is also known to affect blood pressure?
- How do you account for differences in environmental pollutants, a factor that may affect blood pressure?
- Is 100 people in each officially recognized country a large enough sampling frame?

Random Sampling

When we want to analyze something in a large population and get an unbiased cross-section of that population, *random sampling* can be used. In order to ensure that a sampling process is random (or as close as we can get), we use sets, lists, or tables, of so-called *random numbers*.

A RANDOM SAMPLING FRAME

Think of the set T of all the telephone numbers in the United States of America (USA) in which the last two digits are 8 and 5. In theory, any element t of T can be expressed in a format like this:

$$t = abc\text{-}def\text{-}gh85$$

where each value a through h is some digit from the set $\{0, 1, 2, 3, 4, 5, 6, 7, 8, 9\}$. The dashes are not minus signs; they're included to separate number groups, as is customarily done in USA telephone numbers.

If you're familiar with the format of USA phone numbers, you'll know that the first three digits a, b, and c together form the *area code*, and the next three digits d, e, and f represent the *dialing prefix*. Some of the values generated in this way aren't valid USA phone numbers. For example, the format above implies that there can be 1000 area codes, but there aren't that many area codes in the set of valid USA phone numbers. If a randomly chosen phone number isn't a valid USA phone number, we agree to reject it. So all we need to do is generate random digits for each value of a through h in the following generalized number:

$$a,bcd,efg,h85$$

where h represents the "100s digit," g represents the "1000s digit," f represents the "10,000s digit," and so on up to a, which represents the "1,000,000,000s digit." We can pick out strings eight digits long from a random-digit list, and plug them in repeatedly as values a through h to get 10-digit phone numbers ending in 85.

SMALLER AND SMALLER

In the above scenario, we're confronted with a gigantic sampling frame. In reality, the number of elements is smaller than the maximum possible, because many of the telephone numbers derived by random-number generation are not assigned to anybody.

Suppose we conduct an experiment that requires us to get a random sampling of telephone numbers in the USA. By generating samples on the above basis, that is, picking out those that end in the digits 85, we're off to a good start. It is reasonable to think that this sampling frame is an unbiased cross-section of all the phone numbers in the USA. But we'll want to use a smaller sampling frame.

What will happen if we go through a list of all the valid area codes in the USA, and throw out sequences *abc* that don't represent valid area codes? This still leaves us with a sampling frame larger than the set of all assigned numbers in the USA. How about allowing any area code, valid or not, but insisting that the number *ab,cde,fgh* be divisible by 7? Within the set of all possible such numbers, we would find the set of all numbers that produce a connection when dialed, that is, the set of all "actual numbers" (Fig. 5-2). Once we decided on a sampling frame, we could get busy with our research.

REPLACEMENT OR NOT?

Just after an element of a set is sampled, it can be left in the set, or else it can be removed. If the element is left in the set so it can be sampled again, the process is called *sampling with replacement*. If the element is removed so it can't be sampled again, the process is called *sampling without replacement*.

For a finite sample set, if the experiment is continued long enough the set will eventually be exhausted. Table 5-1A shows how this works if the initial sample contains 10 elements, in this case the first 10 letters of the English alphabet. This is presumably what would happen, on a much larger scale, in the hypothetical telephone-number experiment described above. The experimenters would not likely want to count any particular telephone number twice, because that would bias the experiment in favor of the repeated numbers.

If the sample set is infinite – the set of all points on a length of utility cable, for example, or the set of all geographical locations on the surface of the earth – the size of the sample set does not decrease, even if replacement is not done. An infinite set is inexhaustible if its elements are picked out only one at a time.

If the elements of a sample set are replaced after sampling, the size of the sample set remains constant, whether it is finite or infinite. Table 5-1B shows how this works with the first 10 letters of the English alphabet. Note that

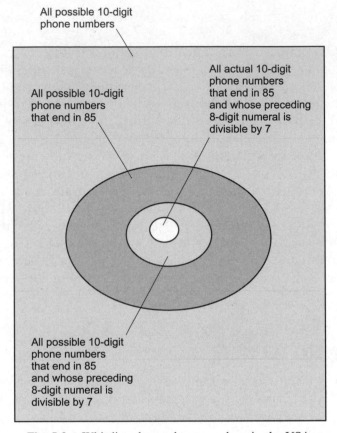

All possible 10-digit
phone numbers

All actual 10-digit
phone numbers
that end in 85
and whose preceding
8-digit numeral is
divisible by 7

All possible 10-digit
phone numbers
that end in 85

All possible 10-digit
phone numbers
that end in 85
and whose preceding
8-digit numeral is
divisible by 7

Fig. 5-2.   Whittling down phone numbers in the USA.

once an element has been replaced in a finite sample set, that element may
(and almost certainly will) be sampled more than once during the course of
the experiment. This happens in the scenario shown by Table 5-1B.

PROBLEM 5-3
Here's a little theoretical problem that you might find fun. Think of it as a
"brain buster." Suppose a sample set is infinite. Consider the set of all
rational numbers between 0 and 1 inclusive. One-by-one sampling can be
done forever from such a set, whether the sampled elements are replaced
or not. If sampling is done at random and without replacement, a given
number will never turn up twice during the course of the experiment, no
matter how long the experiment continues. But suppose sampling is done
with replacement. Can any given number turn up twice in that case?

Table 5-1A When sampling is done without replacement, the size of the sample set decreases and eventually declines to zero. (Read down for experiment progress.)

Before sampling	Element	After sampling
{abcdefghij}	f	{abcdeghij}
{abcdeghij}	d	{abceghij}
{abceghij}	c	{abeghij}
{abeghij}	g	{abehij}
{abehij}	a	{behij}
{behij}	h	{beij}
{beij}	i	{bej}
{bej}	j	{be}
{be}	e	{b}
{b}	b	$\varnothing$

SOLUTION 5-3

If you calculate the probability of any given number turning up after it has been identified, you find that this probability is equal to "one divided by infinity." Presumably this means that the probability is 0, because it is reasonable to suppose that $1/\infty = 0$. (The sideways 8 represents infinity.) This implies that any specific number can never turn up again once it has been selected from an infinite set, even if you replace it after it has been picked.

But wait! The above argument also implies that any specific number between 0 and 1 can never turn up *even once* in a random selection process! Suppose you pick a number between 0 and 1 out of the "clear blue." The probability of this number coming up in your experiment is not equal to 0, because you picked it. This isn't a random number, because your mind is, of course, biased. But it is a specific number. Now imagine a machine that generates truly random numbers between 0 and 1. You push the button,

Table 5-1B When sampling is done with replacement, the size of the sample set stays the same, and some samples may be taken repeatedly. (Read down for experiment progress.)

Before sampling	Element	After sampling
{abcdefghij}	f	{abcdefghij}
{abcdefghij}	d	{abcdefghij}
{abcdefghij}	c	{abcdefghij}
{abcdefghij}	d	{abcdefghij}
{abcdefghij}	a	{abcdefghij}
{abcdefghij}	h	{abcdefghij}
{abcdefghij}	i	{abcdefghij}
{abcdefghij}	i	{abcdefghij}
{abcdefghij}	e	{abcdefghij}
{abcdefghij}	b	{abcdefghij}
{abcdefghij}	a	{abcdefghij}
{abcdefghij}	d	{abcdefghij}
{abcdefghij}	e	{abcdefghij}
{abcdefghij}	h	{abcdefghij}
{abcdefghij}	f	{abcdefghij}
↓	↓	↓

and out comes a number. What is the probability that it's the specific number you have picked? The probability is 0, because the number you happened to choose is only one of an infinite number of numbers between 0 and 1. The machine might choose any one of an infinitude of numbers, and the chance that it's the one you've thought of is therefore $1/\infty$. Therefore, the random-number machine cannot output any specific number.

Can you resolve this apparent paradox?

MINIMIZING ERROR

When conducting a real-world statistical experiment, errors are inevitable. But there are ways in which error can be kept to a minimum. It's important that all experiments be well conceived and well executed. There are various ways in which an experiment can be flawed. The most common sources of this type of error, which we might call *experimental defect error*, include:

- a sample that is not large enough
- a sample that is biased
- replacing elements when they should not be replaced
- failing to replace elements when they should be replaced
- failing to notice and compensate for factors that can bias the results
- attempting to compensate for factors that don't have any real effect
- sloppy measurement of quantities in the sampling process

Improper tallying of the results of a poll or election is a good example of sloppy measurement. If people do not observe the results correctly even though the machines are working the way they ought, it is human error. An election is an example of a digital process. A voter casts either a "yes" (logic 1) or "no" (logic 0) for each candidate.

Measurement error can occur because of limitations in analog hardware. Suppose we want to determine the average current consumed by commercially manufactured 550-watt sodium-vapor lamps when they are operated with only 90 volts rather than the usual 120 volts. We need an alternating-current (AC) ammeter (current-measuring meter) in order to conduct such an experiment. If the ammeter is defective, the results will be inaccurate. No ammeter is perfectly accurate, and with analog ammeters, there is the additional human-error problem of *visual interpolation*. Figure 5-3 shows what an analog AC ammeter might read if placed in the line in series with a high-wattage, 120-volt lamp operating at only 90 volts.

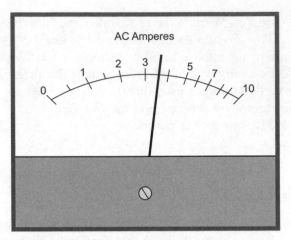

Fig. 5-3. Visual interpolation of an analog meter reading is always subject to error. Illustration for Problems 5-4 through 5-6.

PROBLEM 5-4

Interpolate the reading of the ammeter in Fig. 5-3 to the nearest tenth of an ampere, and to the nearest thousandth of an ampere.

SOLUTION 5-4

Note that the divisions get closer together as the value increases (toward the right). Based on this, and using some common sense, a good estimate of the reading is 3.6 amperes, accurate to the nearest tenth of an ampere.

It is impossible to visually interpolate this ammeter reading to the nearest thousandth of an ampere. We would need a better ammeter in order to do that.

PROBLEM 5-5

How accurately can readings of the ammeter shown in Fig. 5-3 be visually interpolated?

SOLUTION 5-5

This is a matter of opinion, but it's a good bet that experienced meter-readers would say that interpolation to the nearest 0.1 ampere is the limit near the upper end of the scale, and interpolation to the nearest 0.05 ampere is the limit near the lower end. Therefore, we can say that the visual interpolation error is about ±0.05 ampere (plus-or-minus 1/20 of an ampere) near the upper end of the scale, and ±0.025 ampere (plus-or-minus 1/40 of an ampere) near the lower end.

PROBLEM 5-6

Suppose that, in addition to the visual interpolation error, the manufacturer of the ammeter shown in Fig. 5-3 tells us that we can expect a hardware error of up to ±10% of full scale. If that is the case, what is the best accuracy we can expect at the upper end of the scale?

SOLUTION 5-6

A full-scale reading on this ammeter is 10 amperes. So at the upper end of the scale, this instrument could be off by as much as ±(10% × 10), or ±1 ampere. Adding this to the visual interpolation error of ±0.05 ampere, we get a total measurement error of up to ±1.05 ampere at the upper end of the scale.

Estimation

In the real world, characteristics such as the mean and standard deviation can only be approximated on the basis of experimentation. Such approximation is called *estimation*.

ESTIMATING THE MEAN

Imagine there are millions of 550-watt sodium-vapor lamps in the world. Think about the set of all such lamps designed to operate at 120 volts AC, the standard household utility voltage in the USA. Imagine we connect each and every one of these bulbs to a 90-volt AC source, and measure the current each bulb draws at this below-normal voltage. Suppose we get current readings of around 3.6 amperes, as the ammeter in Fig. 5-3 shows. After rejecting obviously defective bulbs, each bulb we test produces a slightly different reading on the ammeter because of imprecision in the production process, just as jelly beans in a bag of candy vary slightly in size.

We do not have the time or resources to test millions of bulbs. So, suppose we select 1000 bulbs at random and test them. We obtain a high-precision digital AC ammeter that can resolve current readings down to the thousandth of an ampere, and we obtain a power supply that is close enough to 90 volts to be considered exact. In this way we get, in effect, a "perfect lab," with measurement equipment that does its job to perfection and eliminates human interpolation error. Variations in current readings therefore represent actual differences in the current drawn by different lamps. We plot the results of the experiment as a graph, smooth it out with curve fitting, and end up with a normal distribution such as the one shown in Fig. 5-4.

Suppose we average all the current measurements and come up with 3.600 amperes, accurate to the nearest thousandth of an ampere. This is an *estimate of the mean* current drawn by all the 550-watt, 120-volt bulbs in the world when they are subjected to 90 volts. This is only an estimate, not the *true mean*, because we have tested only a small sample of the bulbs, not the whole population. The estimate could be made more accurate by testing more bulbs (say 10,000 instead of 1000). Time and money could be saved by testing fewer bulbs (say 100 instead of 1000), but this would produce a less accurate estimate. But we could never claim to know the true mean current drawn by this particular type of bulb at 90 volts, unless we tested every single one of them in the whole world.

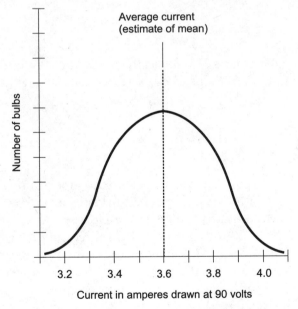

Fig. 5-4.   Estimating the mean. The accuracy improves as the size of the sample set increases.

ESTIMATING THE STANDARD DEVIATION

In the above imaginary situation, we can say that there is a true distribution representing a graphical plot of the current drawn by each and every one of the 120-volt, 550-watt bulbs in the world when subjected to 90 volts. The fact that we lack the resources to find it (we can't test every bulb in the world) does not mean that it does not exist. As the size of the sample increases, Fig. 5-4 approaches the true distribution, and the average current (estimate of the mean current) approaches the true mean current.

True distributions can exist even for populations so large that they can be considered infinite. For example, suppose we want to estimate the mean power output of the stars in the known universe! All the kings, presidents, emperors, generals, judges, sheriffs, professors, astronomers, mathematicians, and statisticians on earth, with all the money in all the economies of humankind, cannot obtain an actual figure for the mean power output of all the stars in the universe. The cosmos is too vast; the number of stars too large. But no one can say that the set of all stars in the known universe does not exist! Thus, the true mean power output of all the stars in the universe is a real thing too.

The rules concerning estimation accuracy that apply to finite populations also apply to infinite populations. As the size of the sample set increases, the accuracy of the estimation improves. As the sample set becomes gigantic, the estimated value approaches the true value.

The mean current drawn at 90 volts is not the only characteristic of the light-bulb distribution we can estimate. We can also estimate the standard deviation of the curve. Figure 5-5 illustrates this. We derive the curve by

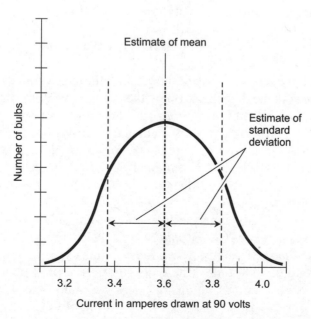

Fig. 5-5.   Estimating the standard deviation. The accuracy improves as the size of the sample set increases.

plotting the points, based on all 1000 individual tests, and then smoothing out the results with curve fitting. Once we have a graph of the curve, we can use a computer to calculate the standard deviation.

From Fig. 5-5, it appears that the standard deviation, σ, is approximately 0.23 amperes either side of the mean. If we test 10,000 bulbs, we'll get a more accurate estimate of σ. If we test only 100 bulbs, our estimate will not be as accurate.

SAMPLING DISTRIBUTIONS

Here's a problem we haven't yet considered. Suppose, in the bulb-testing scenario, our sample consists of 1000 randomly selected bulbs, and we get the results illustrated by Figs. 5-4 and 5-5. What if we repeat the experiment, again choosing a sample consisting of 1000 randomly selected bulbs? We won't get the same 1000 bulbs as we did the first time, so the results of the experiment will be a little different.

Suppose we do the experiment over and over. We'll get a different set of bulbs every time. The results of each experiment will be almost the same, but they will not be exactly the same. There will be a tiny variation in the estimate of the mean from one experiment to another. Likewise, there will be a tiny variation in the estimate of the standard deviation. This variation from experiment to experiment will be larger if the sample size is smaller (say 100 bulbs), and the variation will be smaller if the sample size is larger (say 10,000 bulbs).

Imagine that we repeat the experiment indefinitely, estimating the mean again and again. As we do this and plot the results, we obtain a distribution that shows how the mean varies from sample to sample. Figure 5-6 shows what this curve might look like. It is a normal distribution, but its values are much more closely clustered around 3.600 amperes. We might also plot a distribution that shows how the standard deviation varies from sample to sample. Again we get a normal distribution; its values are closely clustered around 0.23, as shown in Fig. 5-7.

Figures 5-6 and 5-7 are examples of what we call a *sampling distribution*. Figure 5-6 shows a *sampling distribution of means*. Figure 5-7 illustrates a *sampling distribution of standard deviations*. If our experiments involved the testing of more than 1000 bulbs, these distributions would be more centered (more sharply peaked curves), indicating less variability from experiment to experiment. If our experiments involved the testing of fewer than 1000 bulbs, the distributions would be less centered (flatter curves), indicating greater variability from experiment to experiment.

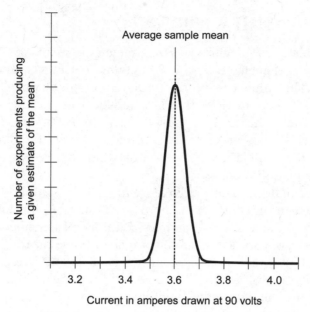

Fig. 5-6. Sampling distribution of mean.

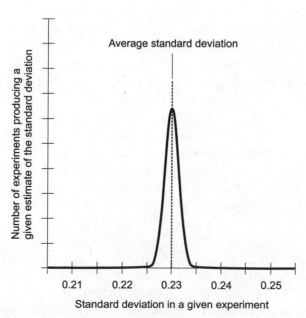

Fig. 5-7. Sampling distribution of standard deviation.

THE CENTRAL LIMIT THEOREM

Imagine a population P in which some characteristic x can vary from element (or individual) to element. Suppose P contains p elements, and p is a very large number. The value x is plotted on the horizontal axis of a graph, and the number y of individuals with characteristic value x is plotted on the vertical axis. The result is a statistical distribution. Maybe it's a normal distribution (bell-shaped and symmetrical), and maybe not. The number of elements p in the population P is so large that it's easiest to render the distribution as a smooth curve.

Now imagine that we choose a large number, k, of samples from P. Each sample represents a different random cross-section of P, but all the samples are the same size. Each of the k samples contains n elements, where $n < p$. We find the mean of each sample and compile all these means into a set $\{\mu_1, \mu_2, \mu_3, \ldots, \mu_k\}$. We then plot these means on a graph. We end up with a sampling distribution of means. We've been through this discussion with the example involving the light bulbs, and now we're stating it in general terms. We're repeating this concept because it leads to something important known as the *Central Limit Theorem*.

According to the first part of the Central Limit Theorem, the sampling distribution of means is a normal distribution if the distribution for P is normal. If the distribution for P is not normal, then the sampling distribution of means approaches a normal distribution as the sample size n increases. Even if the distribution for P is highly skewed (asymmetrical), any sampling distribution of means is more nearly normal than the distribution for P. It turns out that if $n \geq 30$, then even if the distribution for P is highly skewed and p is gigantic, for all practical purposes the sampling distribution of means is a normal distribution.

The second part of the Central Limit Theorem concerns the standard deviation of the sampling distribution of means. Let σ be the standard deviation of the distribution for some population P. Let n be the size of the samples of P for which a sampling distribution of means is determined. Then the standard deviation of the sampling distribution of means, more often called the *standard error of the mean (SE)*, can be found with the following formula:

$$SE \approx \sigma/(n^{1/2})$$

That is, SE is approximately equal to the standard deviation of the distribution for P, divided by the square root of the number of elements in each sample. If the distribution for P is normal, or if $n \geq 30$, then we can consider the formula exact:

$$SE = \sigma/(n^{1/2})$$

From this it can be seen that as the value of n increases, the value of SE decreases. This reflects the fact that large samples, in general, produce more accurate experimental results than small samples. This holds true up to about $n = 30$, beyond which there is essentially no further accuracy to be gained.

Confidence Intervals

A distribution can provide us with generalized data about populations, but it doesn't tell us much about any of the individuals in the population. *Confidence intervals* give us a better clue as to what we can expect from individual elements taken at random from a population.

THE SCENARIO

Suppose we live in a remote scientific research outpost where the electric generators only produce 90 volts. The reason for the low voltage is the fact that the generators are old, inefficient, and too small for the number of people assigned to the station. Funding has been slashed, so new generators can't be purchased. There are too many people and not enough energy. (Sound familiar?)

We find it necessary to keep the compound well-lit at night, regardless of whatever other sacrifices we have to make. So we need bright light bulbs. We have obtained data sheets for 550-watt sodium-vapor lamps designed for 120-volt circuits, and these sheets tell us how much current we can expect the lamps to draw at various voltages. Suppose we have obtained the graph in Fig. 5-8, so we have a good idea of how much current each bulb will draw from our generators that produce 90 volts. The estimate of the mean, μ^*, is 3.600 amperes. There are some bulbs that draw a little more than 3.600 amperes, and there are some that draw a little less. A tiny proportion of the bulbs draw a lot more or less current than average.

If we pick a bulb at random, which is usually what happens when anybody buys a single item from a large inventory, how confident can we be that the current our lamp draws will be within a certain range either side of 3.600 amperes?

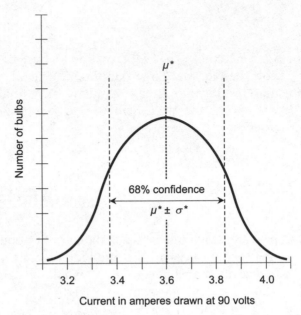

Fig. 5-8.   The 68% confidence interval spans values in the range $\mu^* \pm \sigma^*$.

68% CONFIDENCE INTERVAL

Imagine that our data sheets tell us the standard deviation of the distribution shown in Fig. 5-8 is 0.230 amperes. According to the empirical rule, which we learned about in Chapter 3, 68% of the elements in a sample have a parameter that falls within one standard deviation ($\pm\sigma$) of the mean μ for that parameter in a normal distribution. We don't know the actual standard deviation σ or the actual mean μ for the lamps in our situation, but we have estimates μ^* and σ^* that we can use to get a good approximation of a *confidence interval*.

In our situation, the parameter is the current drawn at 90 volts. Therefore, 68% of the bulbs can be expected to draw current that falls in a range equal to the estimate of the mean plus or minus one standard deviation ($\mu^* \pm \sigma^*$). In Fig. 5-8, this range is 3.370 amperes to 3.830 amperes. It is a *68% confidence interval* because, if we select a single bulb, we can be 68% confident that it will draw current in the range between 3.370 and 3.830 amperes when we hook it up to our sputtering, antiquated 90-volt generator.

95% CONFIDENCE INTERVAL

According to the empirical rule, 95% of the elements have a parameter that falls within two standard deviations of the mean for that parameter in a normal distribution. Again, we don't know the actual mean and standard deviation. We only have estimates of them, because the data is not based on tests of all the bulbs of this type that exist in the world. But we can use the estimates to get a good idea of the *95% confidence interval*.

In our research-outpost scenario, 95% of the bulbs can be expected to draw current that falls in a range equal to the estimate of the mean plus or minus two standard deviations ($\mu^* \pm 2\sigma^*$). In Fig. 5-9, this range is 3.140 amperes to 4.060 amperes.

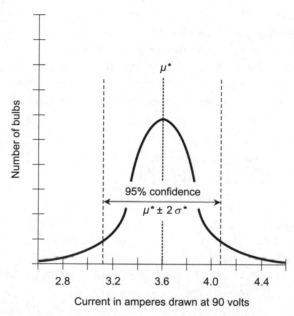

Fig. 5-9.   The 95% confidence interval spans values in the range $\mu^* \pm 2\sigma^*$.

The 95% confidence interval is often quoted in real-world situations. You may hear that "there's a 95% chance that Ms. X will survive her case of cancer for more than one year," or that "there is a 95% probability that the eyewall of Hurricane Y will not strike Miami." If such confidence statements are based on historical data, we can regard them as reflections of truth. But the way we see them depends on where we are. If you have an inoperable malignant tumor, or if you live in Miami and are watching a hurricane prowling the Bahamas, you may take some issue with the use of the word "confidence" when talking about your future. Statistical data can look a lot different to us when our own lives are on the line, as compared to the situation where we are in some laboratory measuring the currents drawn by light bulbs.

WHEN TO BE SKEPTICAL

In some situations, the availability of statistical data can affect the very event the data is designed to analyze or predict. Cancer and hurricanes don't care about polls, but people do!

If you hear, for example, that there is a "95% chance that Dr. J will beat Mr. H in the next local mayoral race," take it with a dose of skepticism. There are inherent problems with this type of analysis, because people's reactions to the publication of predictive statistics can affect the actual event. If broadcast extensively by the local media, a statement suggesting that Dr. J has the election "already won" could cause overconfident supporters of Dr. J to stay home on election day, while those who favor Mr. H go to the polls in greater numbers than they would have if the data had not been broadcast. Or it might have the opposite effect, causing supporters of Mr. H to stay home because they believe they'll be wasting their time going to an election their candidate is almost certain to lose.

99.7% CONFIDENCE INTERVAL

The empirical rule also states that in a normal distribution, 99.7% of the elements in a sample have a parameter that falls within three standard deviations of the mean for that parameter. From this fact we can develop an estimate of the *99.7% confidence interval*.

In our situation, 99.7% of the bulbs can be expected to draw current that falls in a range equal to the estimate of the mean plus or minus three standard deviations ($\mu^* \pm 3\sigma^*$). In Fig. 5-10, this range is 2.910 amperes to 4.290 amperes.

c% CONFIDENCE INTERVAL

We can obtain any confidence interval we want, within reason, from a distribution when we have good estimates of the mean and standard deviation (Fig. 5-11). The width of the confidence interval, specified as a percentage c, is related to the number of standard deviations x either side of the mean in a normal distribution. This relationship takes the form of a function of x versus c.

When graphed for values of c ranging upwards of 50%, the function of x versus c for a normal distribution looks like the curve shown in Fig. 5-12. The curve "blows up" at $c = 100\%$.

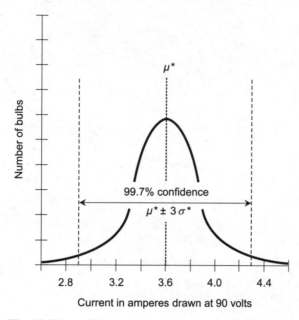

Fig. 5-10.   The 99.7% confidence interval spans values in the range $\mu^* \pm 3\sigma^*$.

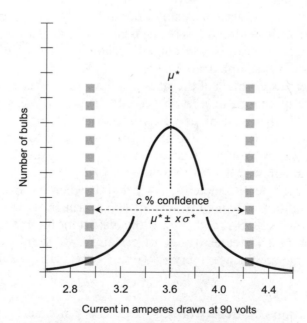

Fig. 5-11.   The $c\%$ confidence interval spans values in the range $\mu^* \pm x\sigma^*$.

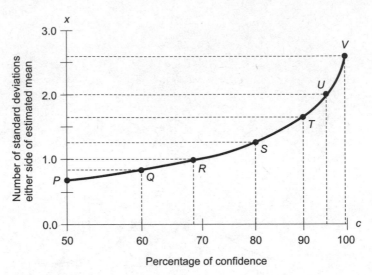

Fig. 5-12. Graph of x as a function of c. Points are approximately: $P = (50\%, 0.67)$, $Q = (60\%, 0.84)$, $R = (68\%, 1.00)$, $S = (80\%, 1.28)$, $T = (90\%, 1.64)$, $U = (95\%, 2.00)$, and $V = (99\%, 2.58)$.

INEXACTNESS AND IMPOSSIBILITY

The foregoing calculations are never exact. There are two reasons for this.

First, unless the population is small enough so we can test every single element, we can only get estimates of the mean and standard deviation, never the actual values. This can be overcome by using good experimental practice when we choose our sample frame and/or samples.

Second, when the estimate of the standard deviation σ^* is a sizable fraction of the estimate of the mean μ^*, we get into trouble if we stray too many multiples of σ^* either side of μ^*. This is especially true as the parameter decreases. If we wander too far to the left (below μ^*), we get close to zero and might even stumble into negative territory – for example, "predicting" that we could end up with a light bulb that draws less than no current! Because of this, confidence interval calculations work only when the span of values is a small fraction of the estimate of the mean. This is true in the cases represented above and by Figs. 5-8, 5-9, and 5-10. If the distribution were much flatter, or if we wanted a much greater degree of certainty, we would not be able to specify such large confidence intervals without modifying the formulas.

PROBLEM 5-7

Suppose you set up 100 archery targets, each target one meter (1 m) in radius, and have thousands of people shoot millions of arrows at these targets from

10 m away. Each time a person shoots an arrow at the target, the radius r at which the arrow hits, measured to the nearest millimeter (mm) from the exact center point P of the bull's eye, is recorded and is fed into a computer. If an arrow hits to the left of a vertical line L through P, the radius is given a negative value; if an arrow hits to the right of L, the radius is given a positive value as shown in Fig. 5-13. As a result, you get a normal distribution in which the mean value, μ, of r is 0. Suppose you use a computer to plot this distribution, with r on the horizontal axis and the number of shots for each value of r (to the nearest millimeter) on the vertical axis. Suppose you run a program to evaluate this curve, and discover that the standard deviation, σ, of the distribution is 150 mm. This is not just an estimate, but is an actual value, because you record all of the shots.

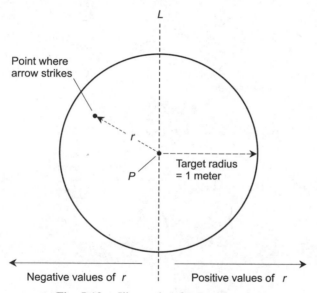

Fig. 5-13. Illustration for Problem 5-7.

If you take a person at random from the people who have taken part at the experiment and have him or her shoot a single arrow at a target from 10 m away, what is the radius of the 68% confidence interval? The 95% confidence interval? The 99.7% confidence interval? What do these values mean? Assume there is no wind or other effect that would interfere with the experiment.

SOLUTION 5-7
The radius of the 68% confidence interval is equal to σ, or 150 mm. This means that we can be 68% confident that our subject's shot will land within

150 mm of the center point, P. The radius of the 95% confidence interval is 2σ, or 300 mm. This means we can be 95% sure that the arrow will land within 300 mm of P. The 99.7% confidence interval is equal to 3σ, or 450 mm. This means we can be 99.7% sure that the arrow will land within 450 mm of P.

PROBLEM 5-8
Draw a graph of the distribution resulting from the experiment described in Problem 5-7, showing the 68%, 95%, and 99.7% confidence intervals.

SOLUTION 5-8
You should get a curve that looks like the one shown in Fig. 5-14.

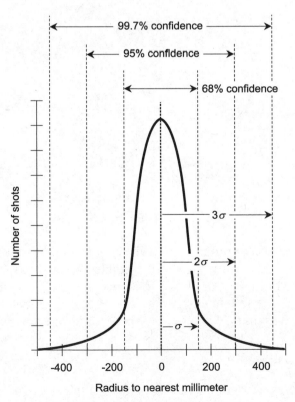

Fig. 5-14. Illustration for Problem 5-8.

Quiz

Refer to the text in this chapter if necessary. A good score is 8 correct. Answers are in the back of the book.

1. Primary source data is
   (a) collected by someone else and then used by a statistician
   (b) collected and used by a statistician
   (c) another expression for a confidence interval
   (d) another expression for a parameter

2. Estimation gives a result that is almost always
   (a) an approximation
   (b) exact
   (c) primary source data
   (d) secondary source data

3. In order to get the best possible estimate of the standard deviation for a certain characteristic of an infinite population, we should
   (a) make the sample set as large as we can
   (b) use the smallest sample set we can get away with
   (c) test all the elements in the population
   (d) test 1000 elements at random

4. As the width of the confidence interval in a normal distribution decreases, the probability of a randomly selected element falling within that interval
   (a) decreases
   (b) does not change
   (c) increases
   (d) approaches a value of 68%

5. If items are repeatedly selected at random, with replacement, from a set of 750,000 elements:
   (a) the size of the set does not change
   (b) the size of the set increases
   (c) the size of the set decreases
   (d) the size of the set becomes zero

6. Suppose you have a normal distribution, and you have good estimates of the mean and standard deviation. Which of the following cannot be determined on the basis of this information?
   (a) The 60% confidence interval.

(b) The 80% confidence interval.

(c) The 90% confidence interval.

(d) All of the above (a), (b), and (c) can be determined on the basis of this information.

7. Which of the following is not necessarily a problem in an experiment?

(a) A biased sample.

(b) The replacement of elements when they should not be replaced.

(c) The use of a random number table.

(d) Failing to compensate for factors that can skew the results.

8. A sampling frame is

(a) a subset of a sample

(b) a subset of a population

(c) an element of a sample

(d) an element of a population

9. Suppose you want to determine the percentage of people in your state who smoke more than five cigarettes a day. Which of the following samples should you expect would be the best (least biased)?

(a) All the people in the state over age 95.

(b) All the hospital patients in the state.

(c) All the elementary-school students in the state.

(d) Every person in the state who was born on June 17.

10. When sampling is done without replacement in an infinite set:

(a) the size of the set does not change

(b) the size of the set increases

(c) the size of the set decreases

(d) the size of the set becomes zero

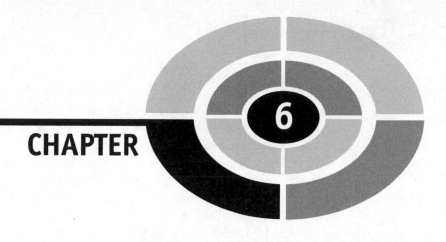

Hypotheses, Prediction, and Regression

In this chapter, we'll look at some of the techniques statisticians use to conduct experiments, interpret data, and draw conclusions.

Assumptions and Testing

A *hypothesis* is a supposition. Some hypotheses are based on experience, some on reason, and some on opinion. A hypothesis can be "picked out of the blue" to see what would happen (or be likely to happen) under certain circumstances. Often, the truth or falsity of a hypothesis can greatly affect the outcome of a statistical experiment.

IT'S AN ASSUMPTION

A hypothesis is always an assumption. Often it takes the form of a prediction. Maybe it is correct; maybe not. Maybe it will prove true, maybe not. Maybe we will never know one way or the other.

When data is collected in a real-world experiment, there is always some error. This error can result from instrument hardware imperfections, limitations of the human senses in reading instrument displays, and sometimes plain carelessness. But in some experiments or situations, there is another source of potential error: large portions of vital data are missing. Maybe it is missing because it can't be obtained. Maybe it has been omitted intentionally by someone who hopes to influence the outcome of the experiment. In such a situation, we might have to make an "educated guess" concerning the missing data. This "guess" can take the form of a hypothesis.

Imagine a major hurricane named *Emma* churning in the North Atlantic. Suppose you live in Wilmington, Delaware, and your house is on the shore of the Delaware River. Does the hurricane pose a threat to you? If so, when should you expect the danger to be greatest? How serious will the situation be if the hurricane strikes? The answers to these questions are dependent variables. They depend on several factors. Some of the factors can be observed and forecast easily and with accuracy. Some of the factors are difficult to observe or forecast, or can be only roughly approximated.

The best the meteorologist and statistician can do in this kind of situation is formulate a graph that shows the probability that the hurricane will follow a path between two certain limits. An example of such a plot, showing the situation for the hypothetical Hurricane Emma, is shown in Fig. 6-1. The probability that the storm will strike between two limiting paths is indicated

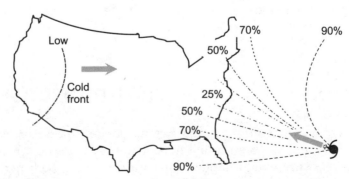

Fig. 6-1. To predict where a hurricane will go, we must formulate hypotheses as well as obtain hard data.

by the percentage numbers. This is why the values increase as the limiting paths get farther away, in either direction, from the predicted path, which lies midway between the two dashed lines marked 25%.

MULTIPLE HYPOTHESES

The weather experts use several *models* to make path predictions for Hurricane Emma. Each of these models uses data, obtained from instrument readings and satellite imagery, and processes the data using specialized programs on a supercomputer. The various models "think" in different ways, so they don't all agree. In addition to the computer programs, the hurricane experts use historical data, and also some of their own intuition, to come up with an official storm path forecast and an official storm intensity forecast for the next 24 hours, 2 days, 3 days, and 5 days.

Imagine that, in our hypothetical scenario, all the computer models agree on one thing: Emma, which is a category-5 hurricane (the most violent possible), is going to stay at this level of intensity for the next several days. The weather experts also agree that if Emma strikes land, the event will be remembered by the local residents for a long time. If it goes over Wilmington, the Delaware River will experience massive tidal flooding. You, who live on the riverfront, do not want to be there if and when that occurs. You also want to take every reasonable precaution to protect your property from damage in case the river rises.

Along with the data, there are hypotheses. Imagine that there is a large cyclonic weather system (called a low), with a long trailing cold front, moving from west to east across the USA. A continental low like this can pull a hurricane into or around itself. The hurricane tends to fall into, and then follow, the low, as if the hurricane were a rolling ball and the front were a trough. (This is where the expression "trough" comes from in weather jargon.) But this only happens when a hurricane wanders close enough to get caught up in the wind circulation of the low. If this occurs with Emma, the hurricane will likely be deflected away from the coast, or else make landfall further north than expected. Will the low, currently over the western United States, affect Emma? It's too early to know. So we formulate hypotheses:

- The low crossing North America will move fast, and will interact with Emma before the hurricane reaches land, causing the hurricane to follow a more northerly path than that implied by Fig. 6-1.
- The low crossing North America will stall or dissipate, or will move slowly, and Emma will follow a path near, or to the south of, the one implied by Fig. 6-1.

Imagine that we enter the first hypothesis into the various computer models along with the known data. In effect, we treat the hypothesis as if it were factual data. The computer models create forecasts for us. In the case of the first hypothesis, we might get a forecast map that looks like Fig. 6-2A. Then we enter the second hypothesis into the computer programs. The result might be a forecast map that looks like Fig. 6-2B.

We can make the hypothesis a variable. We can assign various forward speeds to the low-pressure system and the associated front, enter several values into the computer models, and get maps for each value.

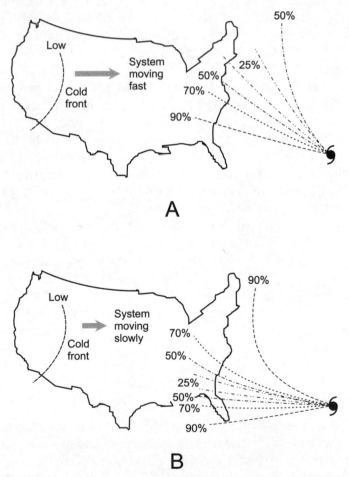

Fig. 6-2. Two hypotheses. At A, path probabilities for a hurricane with fast-moving weather system on continent. At B, path probabilities for a hurricane with slow-moving weather system on continent.

NULL HYPOTHESIS

Suppose we observe the speed of the low-pressure weather system, and its associated cold front, that is crossing the continent. Imagine that we make use of the official weather forecasts to estimate the big system's speed over the next few days. Now suppose that we input this data into a hurricane-forecasting program, and come up with a forecast path for Emma. The program tells us that the *mean path* for Emma will take it across the mid-Atlantic coast of the USA. The computer even generates a graphic illustrating this path (Fig. 6-3).

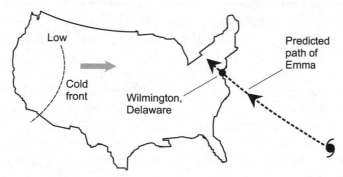

Fig. 6-3. A null hypothesis is a claim or prediction, in this case a forecast path for Hurricane Emma.

We now have a hypothesis concerning the future path of Hurricane Emma. The actual outcome is independent of human control. If we decide to put the hypothesis illustrated by Fig. 6-3 to the test, it is called the *null hypothesis*. A null hypothesis is symbolized H_0 (read "H-null" or "H-nought").

ALTERNATIVE HYPOTHESES

People who lie in the predicted path of Emma, as shown in Fig. 6-3, hope that H_0 is wrong. The *alternative hypothesis*, symbolized H_1, is the proposition that Emma will not follow the path near the one shown in Fig. 6-3. If someone asserts that Emma will go either north or south of the path assumed by H_0, that person proposes a *two-sided alternative*. If someone else claims that Emma will travel north of the path assumed by H_0, that person proposes a *one-sided alternative*. If yet another person proposes that Emma will go south of the path assumed by H_0, it is also a one-sided alternative.

In a situation of this sort, it's possible for a whole crowd of people to come out and propose hypotheses: "I think the storm will hit Washington, DC."

"No, I think it will hit New York City." "I think it will hit somewhere between New York City and Boston, Massachusetts." "You are all wrong. I think it will hit San Francisco, California." "California! You're insane!" The number of possible hypotheses in this scenario is limited only by the number of different opinions that can be obtained. Some are more plausible than others. We can take seriously the notion that the hurricane will strike somewhere between New York and Boston. Most people would reject the hypothesis that Emma will hit San Francisco, although the hypothesis that a person who says so is insane is subject to debate. Maybe she's only joking.

Some people, noting historical data showing that hurricanes almost never strike the mid-Atlantic coast of the USA in a manner such as that given by the computer model and illustrated in Fig. 6-3, claim that the storm will stay further south. Other people think the storm will travel north of the predicted path. There is good reason to believe either of these alternative hypotheses. In the past 100 years or so, the Carolinas and the Northeastern USA have taken direct hits from Atlantic hurricanes more often than has Delaware. There are many different computer programs in use by various government agencies, academic institutions, corporations, and think tanks. Each program produces a slightly different mean path prediction for Emma, given identical input of data. Alternative hypotheses abound. The null hypothesis H_0 is a lonely proposition.

In order to determine whether or not H_0 is correct, the experiment must be carried out. In this instance, that involves no active work on our part (other than getting prepared for the worst), we can only wait and see what happens. Emma will go where Nature steers her.

ICE-CREAM LOVERS

Here's another null/alternative hypothesis situation. Suppose we want to find out what proportion of ice-cream lovers in Canada prefer plain vanilla over all other flavors. Someone makes a claim that 25% of Canadian ice-cream connoisseurs go for vanilla, and 75% prefer some other flavor. This hypothesis is to be tested by conducting a massive survey. It is a null hypothesis, and is labeled H_0.

The simple claim that H_0 is wrong is the basic alternative hypothesis, H_1. If an elderly woman claims that the proportion must be greater than 25%, she asserts a one-sided alternative. If a young boy claims that "nobody in their right mind would like plain vanilla" (so the proportion must be much lower), he also asserts a one-sided alternative. If a woman says she doesn't know whether or not the proportion is really 25%, but is almost certain that the proposition must be wrong one way or the other, then she asserts a two-sided alternative. The experiment in this case consists of carrying out the survey and tallying up the results.

TESTING

In the "USA hurricane scenario," H_0 is almost certain to be rejected after the experiment has taken place. Even though Fig. 6-3 represents the mean path for Emma as determined by a computer program, the probability is low that Emma will actually follow right along this path. If you find this confusing, you can think of it in terms of a double negative. The computers are not saying that Emma is almost sure to follow the path shown in Fig. 6-3. They are telling us that the path shown is the *least unlikely individual path* according to their particular set of data and parameters. (When we talk about the probability that something will or will not occur, we mean the degree to which the forecasters believe it, based on historical and computer data. By clarifying this, we keep ourselves from committing the probability fallacy.)

Similarly, in the "Canada ice-cream scenario," the probability is low that the proportion of vanilla lovers among Canadian ice-cream connoisseurs is exactly 25%. Even if we make this claim, we must be willing to accept that the experiment will almost surely produce results that are a little different, a little more or less than 25%. When we make H_0 in this case, then we are asserting that all other exact proportion figures are less likely than 25%. (When we talk about the probability that something does or does not reflect reality, we mean the degree to which we believe it, based on experience, intuition, or plain guesswork. Again, we don't want to be guilty of the probability fallacy.)

Whenever someone makes a prediction or claim, someone else will refute it. In part, this is human nature. But logic also plays a role. Computer programs for hurricane forecasting get better with each passing year. Methods of conducting statistical surveys about all subjects, including people's ice-cream flavor preferences, also improve.

If a group of meteorologists comes up with a new computer program that says Hurricane Emma will pass over New York City instead of Wilmington, then the output of that program constitutes evidence against H_0 in the "USA hurricane scenario." If someone produces the results of a survey showing that only 17% of British ice-cream lovers prefer plain vanilla flavor and only 12% of USA ice-cream lovers prefer it, this might be considered evidence against H_0 in the "Canada ice-cream scenario." The gathering and presentation of data supporting or refuting a null hypothesis, and the conducting of experiments to figure out the true situation, is called *statistical testing* or *hypothesis testing*.

SPECIES OF ERROR

There are two major ways in which an error can be made when formulating hypotheses. One form of error involves rejecting or denying the potential truth of a null hypothesis, and then having the experiment end up demonstrating that the null hypothesis is true after all. This is sometimes called a *type-1 error*. The other species of error is the exact converse: accepting the null hypothesis and then having the experiment show that it is false. This is called a *type-2 error*.

How likely is either type of error in the "USA hurricane scenario" or the "Canada ice-cream scenario"? These questions can be difficult to answer. It is hard enough to come up with good null hypotheses in the first place. Nevertheless, the chance for error is a good thing to know, because it tells us how seriously we ought to take a null hypothesis. The *level of significance*, symbolized by the lowercase Greek letter alpha (α), is the probability that H_0 will turn out to be true after it has been rejected. This figure can be expressed as a ratio, in which case it is a number between 0 and 1, or as a percentage, in which case it is between 0% and 100%.

What's the Forecast?

Let's revisit the "USA hurricane scenario." The predicted path for Hurricane Emma shown in Fig. 6-3 is a hypothesis, not an absolute. The absolute truth will become known eventually, and a path for Emma will be drawn on a map with certainty, because it will represent history!

THE "PURPLE LINE"

If you live in a hurricane-prone region, perhaps you have logged on to the Internet to get an idea of whether a certain storm threatens you. You look on the tracking/forecast map, and see that the experts have drawn a "purple line" (also known as the "line of doom") going over your town on the map! Does this tell you that the forecasters think the storm is certain to hit you? No. It tells you that the "purple line" represents the mean predicted path based on computer models.

As time passes – that is, as the experiment plays itself out – you will get a better and better idea of how worried you ought to be. If you're the cool-headed scientific sort, you'll go to the Internet sites of the government weather agencies such as the National Hurricane Center, analyze the data for a while, and then make whatever preparations you think are wise. Perhaps you'll decide to take a short vacation to Nashville, Tennessee, and do your statistical analyses of Emma from there.

CONFIDENCE INTERVALS REVISITED

Instead of drawing a single line on a map, indicating a predicted track for Hurricane Emma, it's better to draw path probability maps such as the ones in Figs. 6-1 and 6-2. These maps, in effect, show confidence intervals. As the storm draws closer to the mainland, the confidence intervals narrow. The forecasts are revised. The "purple line" – the mean path of the storm – might shift on the tracking map. (Then again, maybe it won't move at all.) Most hurricane Web sites have strike-probability maps that are more informative than the path-prediction maps.

PROBABILITY DEPENDS ON RANGE

Imagine that a couple of days pass, Emma has moved closer to the mainland, and the forecasters are still predicting a mean path that takes the center of Emma over Wilmington. The probability lines are more closely spaced than they were two days ago. We can generate a distribution curve that shows the relative danger at various points north and south of the predicted point of landfall (which is actually on the New Jersey coast east of Wilmington). Figure 6-4A is a path probability map, and Fig. 6-4B is an example of a statistical distribution showing the relative danger posed by Emma at various distances from the predicted point of landfall. The vertical axis representing the landfall point (labeled 0 in Fig. 6-4B) does not depict the actual probability of strike. Strike probabilities can be ascertained only within various ranges – lengths of coastline – north and/or south of the predicted point of landfall. Examples are shown in Figs. 6-5A and B.

In the "Canada ice-cream scenario," a similar situation exists. We can draw a distribution curve (Fig. 6-6) that shows the taste inclinations of people, based on the H_0 that 25% of them like plain vanilla ice cream better than any other flavor. Given any range or "margin of error" that is a fixed number of percentage points wide, say $\pm 2\%$ either side of a particular point, H_0 asserts that the greatest area under the curve will be obtained when that

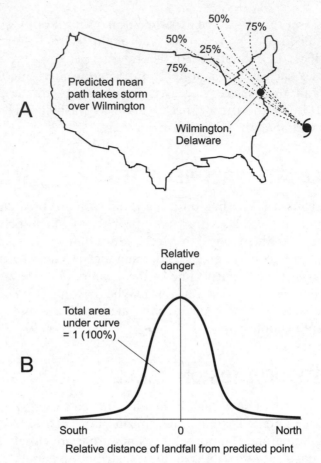

Fig. 6-4. At A, path probabilities for Hurricane Emma as it moves closer to mainland. At B, relative plot of strike danger as a function of distance from predicted point of landfall.

range is centered at 25%. Imagine that H_0 happens to be true. If that is the case, then someone who says "Our survey will show that 23% to 27% of the people prefer vanilla" is more likely to be right than someone who says "Our survey will show that 12% to 16% of the people prefer vanilla." In more general terms, let P be some percentage between 0% and 100%, and let x be a value much smaller than P. Then if someone says "Our survey will show that $P\% \pm x\%$ of the people prefer vanilla," that person is most likely to be right if $P\% = 25\%$. That's where the distribution curve of Fig. 6-6 comes to a peak, and that's where the area under the curve, given any constant horizontal span, is the greatest.

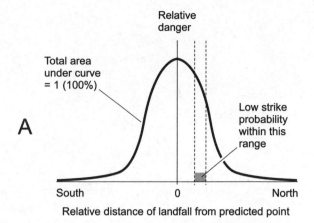

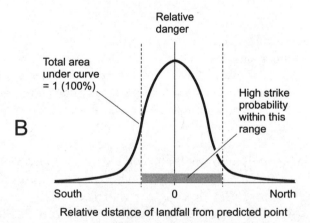

Fig. 6-5. At A, Hurricane Emma is unlikely to strike anywhere within a narrow span of coast. At B, Hurricane Emma is likely to strike somewhere within a wide span of coast.

INFERENCE

The term *inference* refers to any process that is used to draw conclusions on the basis of data and hypotheses. In the simplest sense, inference is the application of reason, common sense, and logic. In statistics, inference requires the application of logic in specialized ways.

We have already seen two tools that can be used for statistical inference: confidence intervals and significance testing. Both of these tools give us numerical output. But ultimately, it is a matter of subjective judgment whether or not we should come to any particular conclusion based on such

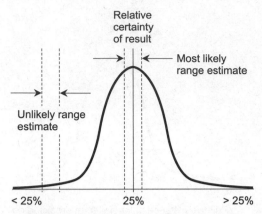

Fig. 6-6. Distribution for the "Canada ice-cream scenario."

data. Sometimes a judgment is easy to make. Sometimes it is difficult. Sometimes inferences can be made and conclusions drawn with a "cool head" because nothing important depends on our decision. Sometimes there are emotional or life-and-death factors that can cloud our judgment. When our judgment is bad, we might make an inference when we should not, or else fail to make an inference when we should.

Consider again the "USA hurricane scenario." If you live on the ocean-front and a hurricane is approaching, what should you do? Board up the windows? Go to a shelter? Find a friend who lives in a house that is better constructed than yours? Get in your car and flee? Statistics can help you decide what to do, but there are no numbers that can define an optimal course of action. No computer can tell you what to do.

In the "Canada ice-cream scenario," suppose we conduct a survey by interviewing 12 people. Three of them (that's 25%) say that they prefer vanilla. Does this mean that H_0, our null hypothesis, is correct? Most people would say no, because 12 people is not a big enough sample. But if we interview 12,000 people (taking care that the ages, ethnic backgrounds, and other factors present an unbiased cross-section of the Canadian population) and 2952 of them say they prefer vanilla, we can reasonably infer that H_0 is valid, because 2952 is 24.6% of 12,000, and that is pretty close to 25%. If 1692 people say they prefer vanilla, we can infer that H_0 is not valid, because 1692 is only 14.1% of 12,000, and that is nowhere near 25%.

How large a sample should we have in order to take the results of our survey seriously? That is a subjective decision. A dozen people is not enough, and 12,000 is plenty; few people will dispute that. But what about 120 people?

Or 240? Or 480? The general rule in a situation like this is to get as large a sample as reasonably possible.

PROBLEM 6-1

Imagine that you are a man and that you live in a town of 1,000,000 people. Recently, you've been seeing a lot of women smoking. You start to suspect that there are more female smokers in your town than male smokers. You discuss this with a friend. Your friend says, "You are wrong. The proportion of female to male smokers is 1:1." You say, "Do you mean to tell me that the number of women smokers in this town is the same as the number of men smokers?" Your friend says, "Yes, or at least pretty close." You counter, "There are far more women smokers than men smokers. I see it every evening. It seems that almost every woman I see has a cigarette in her mouth." Your friend has a quick retort: "That's because you spend a lot of time at night clubs, where the number of women who smoke is out of proportion to the number of women smokers in the general population. Besides that, if I know you, you spend all your time looking at the women, so you haven't noticed whether the men are smoking or not."

Suppose you and your friend decide to conduct an experiment. You intend to prove that there are more female smokers in your town than male smokers. Your friend offers the hypothesis that the number of male and female smokers is the same. What is a good null hypothesis here? What is the accompanying alternative hypothesis? How might we conduct a test to find out who is right?

SOLUTION 6-1

A reasonable null hypothesis, which your friend proposes, is the notion that the ratio of women to men smokers in your town is 1:1, that is, "50-50." Then the alternative hypothesis, which you propose, is that there are considerably more women smokers than men smokers. To conduct a test to find out who is right, you'll have to choose an unbiased sample of the population of your town. The sample will have to consist of an equal number of men and women, and it should be as large as possible. You will have to ask all the subjects whether or not they smoke, and then assume that they're being honest with you.

PROBLEM 6-2

Now imagine, for the sake of argument, that H_0 is in fact true in the above-described scenario. (You don't know it and your friend doesn't know it, because you haven't conducted the survey yet.) You're about to conduct an experiment by taking a supposedly unbiased survey consisting of 100 people, 50 men and 50 women. Draw a simple graph showing the relative probabilities of the null hypothesis being verified, versus either one-sided alternative.

SOLUTION 6-2

The curve is a normal distribution (Fig. 6-7). Of all the possible outcomes, the most likely is a 1:1 split, in which the same number of women as men say they smoke. This doesn't mean that this exact result is certain or even likely; it only means that it is the least unlikely of all the possible outcomes. It's reasonable to suppose that you might get a result of, say, 20 women asserting that they smoke while 22 men say they smoke. But you should be surprised if the survey comes back saying that 40 women say they smoke while only 10 men say so.

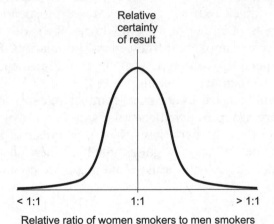

Fig. 6-7. Illustration for Problem 6-2.

PROBLEM 6-3

Name several different possible outcomes of the experiment described above, in which the null hypothesis is apparently verified.

SOLUTION 6-3

Recall that 50 men and 50 women are surveyed. If 20 men say they smoke and 20 women say they do, this suggests the null hypothesis is reasonable. The same goes for ratios such as 15:16 or 25:23.

PROBLEM 6-4

Name two outcomes of the experiment described above, in which the null hypothesis is apparently verified, but in which the results should be highly suspect.

SOLUTION 6-4
If none of the men and none of the women say they smoke, you ought to suspect that a lot of people are lying. Similarly, if all 50 men and all 50 women say they smoke, you should also expect deception. Even ratios of 2:3 or 48:47 would be suspect. (Results such as this might suggest that we conduct other experiments concerning the character of the people in this town.)

Regression

Regression is a way of defining the extent to which two variables are related. Regression can be used in an attempt to predict things, but this can be tricky. The existence of a correlation between variables does not always mean that there is a cause-and-effect link between them.

PAIRED DATA

Imagine two cities, one named Happyton and the other named Blissville. These cities are located far apart on the continent. The prevailing winds and ocean currents produce greatly different temperature and rainfall patterns throughout the year in these two cities. Suppose we are about to move from Happyton to Blissville, and we've been told that Happyton "has soggy summers and dry winters," while in Blissville we should be ready to accept that "the summers are parched and the winters are washouts." We've also been told that the temperature difference between summer and winter is much smaller in Blissville than in Happyton.

We go to the Internet and begin to collect data about the two towns. We find a collection of tables showing the average monthly temperature in degrees Celsius (°C) and the average monthly rainfall in centimeters (cm) for many places throughout the world. Happyton and Blissville are among the cities shown in the tables. Table 6-1A shows the average monthly temperature and rainfall for Happyton as gathered over the past 100 years. Table 6-1B shows the average monthly temperature and rainfall for Blissville over the same period. The data we have found is called *paired data*, because it portrays two variable quantities, temperature and rainfall, side-by-side.

We can get an idea of the summer and winter weather in both towns by scrutinizing the tables. But we can get a more visual-friendly portrayal by making use of bar graphs.

Table 6-1A Average monthly temperature and rainfall for Happyton.

Month	Average temperature, degrees Celsius	Average rainfall, centimeters
January	2.1	0.4
February	3.0	0.2
March	9.2	0.3
April	15.2	2.8
May	20.4	4.2
June	24.9	6.3
July	28.9	7.3
August	27.7	8.5
September	25.0	7.7
October	18.8	3.6
November	10.6	1.7
December	5.3	0.5

PAIRED BAR GRAPHS

Let's graphically compare the average monthly temperature and the average monthly rainfall for Happyton. Figure 6-8A is a *paired bar graph* showing the average monthly temperature and rainfall there. The graph is based on the data from Table 6-1A. The horizontal axis has 12 intervals, each one showing a month of the year. Time is the independent variable. The left-hand vertical scale portrays the average monthly temperatures, and the right-hand vertical scale portrays the average monthly rainfall amounts. Both of these are dependent variables, and are functions of the time of year. The average monthly temperatures are shown by the light gray bars, and the average monthly rainfall amounts are shown by the dark gray bars. It's easy to see from

Table 6-1B Average monthly temperature and rainfall for Blissville.

Month	Average temperature, degrees Celsius	Average rainfall, centimeters
January	10.5	8.1
February	12.2	8.9
March	14.4	6.8
April	15.7	4.2
May	20.5	1.6
June	22.5	0.4
July	23.6	0.2
August	23.7	0.3
September	20.7	0.7
October	19.6	2.4
November	16.7	3.4
December	12.5	5.6

this data that the temperature and rainfall both follow annual patterns. In general, the warmer months are wetter than the cooler months in Happyton.

Now let's make a similar comparison for Blissville. Figure 6-8B is a paired bar graph showing the average monthly temperature and rainfall there, based on the data from Table 6-1B. From this data, we can see that the temperature difference between winter and summer is less pronounced in Blissville than in Happyton. But that's not the main thing that stands out in this bar graph! Note that the rainfall, as a function of the time of year, is much different. The winters in Blissville, especially the months of January and February, are wet. The summers, particularly June, July, and August, get almost no rainfall. The contrast in general climate between Happyton and Blissville is striking. This information is, of course, contained in the tabular data, but it's easier to see by looking at the dual bar graphs.

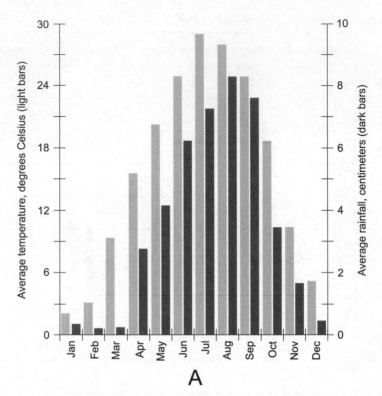

Fig. 6-8A.   Paired bar graph showing the average monthly temperature and rainfall for the hypothetical city of Happyton.

SCATTER PLOTS

When we examine Fig. 6-8A, it appears there is a relationship between temperature and rainfall for the town of Happyton. In general, as the temperature increases, so does the amount of rain. There is also evidently a relationship between temperature and rainfall in Blissville, but it goes in the opposite sense: as the temperature increases, the rainfall decreases. How strong are these relationships? We can draw *scatter plots* to find out.

In Fig. 6-9A, the average monthly rainfall is plotted as a function of the average monthly temperature for Happyton. One point is plotted for each month, based on the data from Table 6-1A. In this graph, the independent variable is the temperature, not the time of the year. There is a pattern to the arrangement of points. The correlation between temperature and rainfall is positive for Happyton. It is fairly strong, but not extremely so. If there were no correlation (that is, if the correlation were equal to 0), the points would be

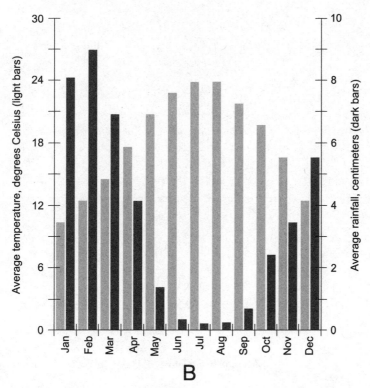

Fig. 6-8B. Paired bar graph showing the average monthly temperature and rainfall for the hypothetical city of Blissville.

randomly scattered all over the graph. But if the correlation were perfect (either +1 or −1), all the points would lie along a straight line.

Figure 6-9B shows a plot of the average monthly rainfall as a function of the average monthly temperature for Blissville. One point is plotted for each month, based on the data from Table 6-1B. As in Fig. 6-9A, temperature is the independent variable. There is a pattern to the arrangement of points here, too. In this case the correlation is negative instead of positive. It is a fairly strong correlation, perhaps a little stronger than the positive correlation for Happyton, because the points seem more nearly lined up. But the correlation is far from perfect.

REGRESSION CURVES

The technique of curve fitting, which we learned about in Chapter 1, can be used to illustrate the relationships among points in scatter plots such as those

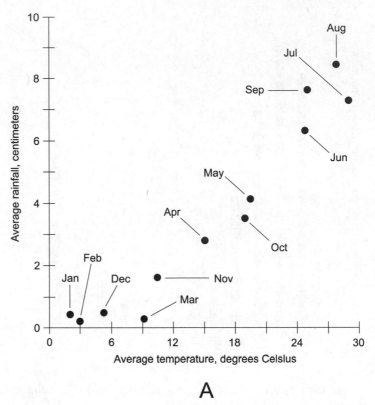

A

Fig. 6-9A. Scatter plot of points showing the average monthly temperature versus the average monthly rainfall for Happyton.

in Figs. 6-9A and B. Examples, based on "intuitive guessing," are shown in Figs. 6-10A and B. Fig. 6.10A shows the same 12 points as those in Fig. 6-9A, representing the average monthly temperature and rainfall amounts for Happyton (without the labels for the months, to avoid cluttering things up). The dashed curve represents an approximation of a smooth function relating the two variables. In Fig. 6-10B, a similar curve-fitting exercise is done to approximate a function relating the average monthly temperature and rainfall for Blissville.

In our hypothetical scenarios, the data shown in Tables 6-1A and B, Figs. 6-8A and B, Figs. 6-9A and B, and Figs. 6-10A and B are all based on records gathered over 100 years. Suppose that we had access to records gathered over the past 1000 years instead! Further imagine that, instead of having data averaged by the month, we had data averaged by the week. In these cases we would get gigantic tables, and the bar graphs would be utterly impossible

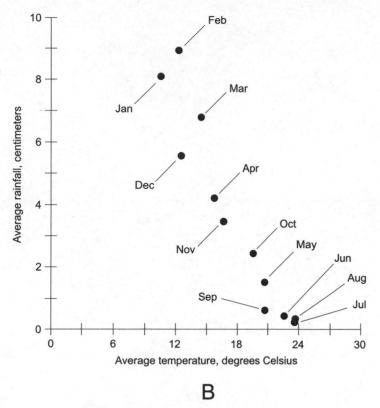

Fig. 6-9B.   Scatter plot of points showing the average monthly temperature versus the average monthly rainfall for Blissville.

to read. But the scatter plots would tell a much more interesting story. Instead of 12 points, each graph would have 52 points, one for each week of the year. It is reasonable to suppose that the points would be much more closely aligned along smooth curves than they are in Figs. 6-9A and B or Figs. 6-10A and B.

LEAST-SQUARES LINES

As you might guess, there is computer software designed to find ideal curves for scatter plots such as those shown in Figs. 6-9A and B. Most high-end scientific graphics packages contain curve-fitting programs. They also contain programs that can find the best overall straight-line fit for the points in any scatter plot where correlation exists. Finding the ideal straight line is easier than finding the ideal smooth curve, although the result is usually less precise.

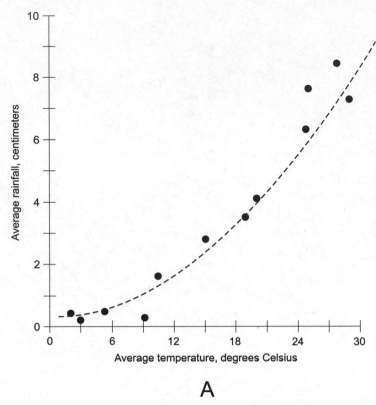

A

Fig. 6-10A. Regression curve relating the average monthly temperature to the average monthly rainfall for Happyton.

Examine Figs. 6-11A and B. These graphs represent the outputs of hypothetical computer programs designed to find the best straight-line approximations of the data from Figs. 6-9A and B, respectively. Suppose that the dashed lines in these graphs represent the best overall straight-line averages of the positions of the points. In that case, then the lines both obey a rule called the *law of least squares.*

Here's how a *least-squares line* is found. Suppose the distance between the dashed line and each of the 12 points in Fig. 6-11A is measured. This gives us a set of 12 distance numbers; call them d_1 through d_{12}. They should all be expressed in the same units, such as millimeters. Square these distance numbers, getting $d_1{}^2$ through $d_{12}{}^2$. Then add these squared numbers together, getting a final sum D. There is one particular straight line for the scatter plot in Fig. 6-11A (or for any scatter plot in which there is a correlation among the points) for which the value of D is minimum. That line is the least-

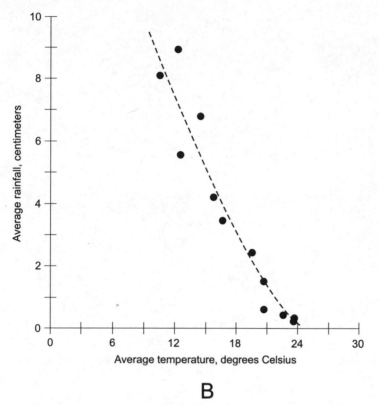

Fig. 6-10B. Regression curve relating the average monthly temperature to the average monthly rainfall for Blissville.

squares line. We can do exactly the same thing for the points and the dashed line in Fig. 6-11B.

Any computer program designed to find the least-squares line for a scatter plot executes the aforementioned calculations and performs an *optimization problem*, quickly calculating the equation of, and displaying, the line that best portrays the overall relationship among the points. Unless the points are randomly scattered or are arranged in some bizarre coincidental fashion (such as a perfect circle, uniformly spaced all the way around), there is always one, and only one, least-squares line for any given scatter plot.

PROBLEM 6-5

Suppose the points in a scatter plot all lie exactly along a straight line so that the correlation is either +1 (as strong, positively, as possible) or −1 (as strong,

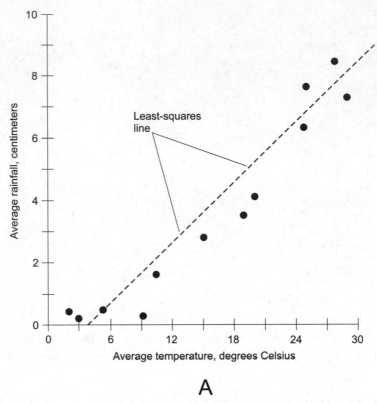

A

Fig. 6-11A. Least-squares line relating the average monthly temperature to the average monthly rainfall for Happyton.

negatively, as possible). Where is the least-squares line in this type of scenario?

SOLUTION 6-5
If all the points in a scatter plot happen to be arranged in a straight line, then that line is the least-squares line.

PROBLEM 6-6
Imagine that the points in a scatter plot are all over the graph, so that the correlation is 0. Where is the least-squares line in this case?

SOLUTION 6-6
When there is no correlation between two variables and the scatter plot shows this by the appearance of randomly placed points, then there is no least-squares line.

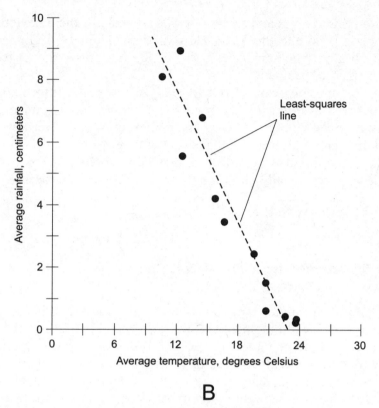

Fig. 6-11B.   Least-squares line relating the average monthly temperature to the average monthly rainfall for Blissville.

PROBLEM 6-7

Imagine the temperature versus rainfall data for the hypothetical towns of Happyton and Blissville, discussed above, has been obtained on a daily basis rather than on a monthly basis. Also suppose that, instead of having been gathered over the past 100 years, the data has been gathered over the past 1,000,000 years. We should expect this would result in scatter plots with points that lie neatly along smooth lines or curves. We might also be tempted to use this data to express the present-day climates of the two towns. Why should we resist that temptation?

SOLUTION 6-7

The "million-year data" literally contains too much information to be useful in the present time. The earth's overall climate, as well as the climate in any particular location, has gone through wild cycles over the past 1,000,000 years. Any climatologist, astronomer, or earth scientist can tell you that.

There have been ice ages and warm interglacial periods; there have been wet and dry periods. While the 1,000,000-year data might be legitimate as it stands, it does not necessarily represent conditions this year, or last year, or over the past 100 years.

Statisticians must be careful not to analyze too much information in a single experiment. Otherwise, the results can be skewed, or can produce a valid answer to the wrong question. The gathering of data over a needlessly large region or an unnecessarily long period of time is a tactic sometimes used by people whose intent is to introduce bias while making it look as if they have done an exceptionally good job of data collection. Beware!

Quiz

Refer to the text in this chapter if necessary. A good score is 8 correct. Answers are in the back of the book.

1. Suppose that in the Happyton/Blissville situation described above, we are told at the last moment that we're not moving to Blissville after all, but instead are going to be relocated to the town of Borington. When we look at the weather data for Borington, we can hardly believe our eyes. The average temperature is 20°C for every month of the year, and the average rainfall is 4.1 cm for every month of the year. When we plot this data on a paired bar graph with coordinate axes like those in Fig. 6-8A or B, we get
   (a) a set of 24 alternating light and dark gray bars that increase steadily in height from left to right
   (b) a set of 24 alternating light and dark gray bars that decrease steadily in height from left to right
   (c) a set of 12 light gray bars that are all the same height, interwoven with a set of 12 dark gray bars that are all the same height
   (d) a set of 24 alternating light and dark gray bars, all 24 of which are the same height

2. Suppose a computer program is used in an attempt to find a least-squares line for a scatter plot. The computer says that no such line exists. Upon visual examination, we can see that the points are spread out over the graph. Evidently the correlation between the two variables
   (a) is between 0 and +1
   (b) is equal to 0
   (c) is between −1 and 0
   (d) is equal to −1

3. In a statistical experiment, error can result from any of the following, *except*
    (a) instrument imperfections
    (b) the use of instruments with greater precision than necessary
    (c) large amounts of missing data
    (d) limitations of the experimenter's ability to read instrument displays

4. The mathematical process of finding the least-squares line in a scatter plot is a sophisticated example of
    (a) curve fitting
    (b) an alternative hypothesis
    (c) a null hypothesis
    (d) strong correlation

5. Suppose we are about to relocate to a small town called Hoodooburg in a sparsely populated part of the western USA. We are curious as to what proportion of the people eat beef steak at least once a week. You make the assumption that the figure is 90%. I disagree, and contend that the proportion is lower than that. Your assumption, which I will try to disprove, is an example of
    (a) a null hypothesis
    (b) a type-1 error
    (c) an alternative hypothesis
    (d) a type-2 error

6. In the scenario of Question 5, my assumption (less than 90% of the people in Hoodooburg eat beef steak at least once a week) is
    (a) a null hypothesis
    (b) a type-1 error
    (c) an alternative hypothesis
    (d) a type-2 error

7. Suppose that we conduct a well-designed, unbiased survey of the population in the scenario of Question 5, and it turns out that you were right and I was wrong. This means I have made
    (a) a null hypothesis
    (b) a type-1 error
    (c) an alternative hypothesis
    (d) a type-2 error

8. Suppose we want to see how blood pressure correlates with age, and we have the medical records of 600 people (which we obtained with

their permission). The clearest way to graphically illustrate this correlation, if there is any, is to put the data in the form of

(a) a table
(b) a null hypothesis
(c) an alternative hypothesis
(d) a scatter plot

9. Examine the scatter plot of Fig. 6-12. Imagine that this graph shows the
relative energy available from sunlight at a number of hypothetical
locations around the world, as determined by actual observations conducted over a period of several years. Then look at the three examples
shown in Fig. 6-13. Which, if any, of the graphs in Fig. 6-13 best
illustrates the least-squares line for the scatter plot of Fig. 6-12?

(a) Figure 6-13A.
(b) Figure 6-13B.
(c) Figure 6-13C.
(d) None of them.

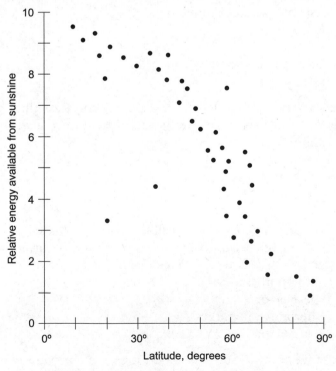

Fig. 6-12. Illustration for Quiz Question 9.

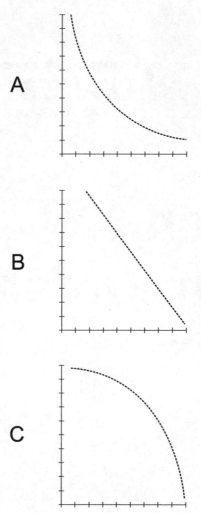

Fig. 6-13.   Illustrations for possible answers to Quiz Question 9. At A, choice (a); at B, choice
(b); at C, choice (c).

10. The use of significance testing is an example of
   (a) an alternative hypothesis
   (b) a null hypothesis
   (c) confidence intervals
   (d) statistical inference

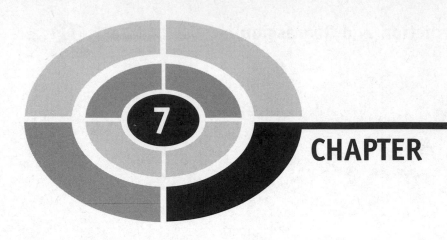

Correlation, Causation, Order, and Chaos

In this chapter, we'll see how cause-and-effect relationships can exist along with correlation. We'll also look at some scenarios in which statistics borders on esoteric disciplines of mathematics, including chaos and the theory of boundaries.

Correlation Principles

We looked at correlation briefly in Chapter 1. Let's examine this phenomenon a little more closely now. When two things are correlated, does one cause the other? Does a third phenomenon cause both? Is there any cause-and-effect relationship at all? People often conclude that there is a cause-and-effect relationship when they see a correlation. But this is not necessarily true.

QUANTITATIVE VERSUS QUALITATIVE

Correlation (often symbolized by the italicized, lowercase letter r) can be numerically defined only between variables that can be quantified. Examples of *quantitative* variables include time, temperature, and average monthly rainfall.

It's possible to *qualitatively* express the correlation between two variables if one or both of them cannot be quantified. But it's not possible to quantitatively express correlation unless both variables and their relationship can be quantified. Even if it seems obvious that two variables are correlated, there is a big difference between saying that, for example, "rudeness and violence are strongly correlated" and "the correlation between rudeness and violence is +0.75." Violence can be quantified on the basis of crime statistics, but rudeness is a more elusive variable to numerically express.

Imagine that a massive social experiment is conducted over a period of years, and researchers come to the conclusion that people develop schizophrenia more often in some geographic regions than in others. Suppose, for example, that there are more people with this disorder living at high elevations in the mountains, where there is lots of snow and the weather is cool all year round, than there are at low elevations near tropical seas, where it rains often and the weather is warm all year. Both of these variables – schizophrenia and environment – are difficult or impossible to quantify. In particular, if you took 100 psychiatrists and asked them to diagnose a person who behaves strangely, you might end up with 40 diagnoses of "schizophrenia," 10 diagnoses of "paranoid psychosis," 15 diagnoses of "depression," 5 diagnoses of "bipolar disorder," 25 diagnoses of "normal but upset," and 5 verdicts of "not enough information to make a diagnosis." While the largest proportion (40%) of the doctors think the person has schizophrenia in this breakdown, that is not even a simple majority. Such a diagnosis is not absolute, such as would be the case with an unmistakable physical ailment such as a brain tumor.

CORRELATION RANGE

The first thing we should know about correlation, as shown or implied by a scatter plot, was suggested earlier in this book. But it's so important that it

bears repetition. Correlation can be expressed as a numerical value r such that the following restriction holds:

$$-1 \leq r \leq +1$$

This means the mathematical correlation can be equal to anything between, and including, −1 and +1. Sometimes percentages are used instead, so the possible range of correlation values, $r_\%$, is as follows:

$$-100\% \leq r_\% \leq +100\%$$

A correlation value of $r = -1$ represents the strongest possible negative correlation; $r = +1$ represents the strongest possible positive correlation. Moderately strong positive correlation might be reflected by a figure of $r = +0.7$; weak negative correlation might show up as $r_\% = -20\%$. A value of $r = 0$ or $r_\% = 0\%$ means there is no correlation at all. Interestingly, the absence of any correlation can be more difficult to prove than the existence of correlation, especially if the number of samples (or points in a scatter plot) is small.

It's impossible for anything to be correlated with anything else to an extent beyond the above limits. If you ever hear anyone talking about two phenomena being correlated by "a factor of −2" or "$r = 150\%$," you know they're wrong. In addition to this, we need to be careful when we say that two effects are correlated to "twice the extent" of two other effects. If two phenomena are correlated by a factor of $r = +0.75$, and someone comes along and tells you that changing the temperature (or some other parameter) will "double the correlation," you know something is wrong because this suggests that the correlation could become $r = +1.50$, an impossibility.

CORRELATION IS LINEAR

There are plenty of computer programs that can calculate correlation numbers based on data input or scatter plots. In this book, we won't get into the actual formulas used to calculate correlation. The formulas are messy and tedious for any but the most oversimplified examples. At this introductory level, it's good enough for you to remember that correlation is a measure of the extent to which the points in a scatter plot are concentrated near the least-squares line.

The key word in correlation determination is the word "line." Correlation in a scatter plot is defined by the nearness of the points to a particular straight line determined from the points on the plot. If points lie along a perfectly straight line, then either $r = -1$ or $r = +1$. The value of r is positive if the values of both variables increase together. The value of r is negative if one value decreases as the other value increases.

Once in a while, you'll see a scatter plot in which all the points lie along a smooth curve, but that curve is not a straight line. This is a special sort of perfection in the relationship between the variables; it indicates that one is a mathematical function of the other. But points on a non-straight curve do not indicate a correlation of either −1 or +1. Figure 7-1A shows a scatter plot in which the correlation is +1. Figure 7-1B shows a scatter plot in which the correlation is perfect in the sense that the points lie along a smooth curve, but in fact the correlation is much less than +1.

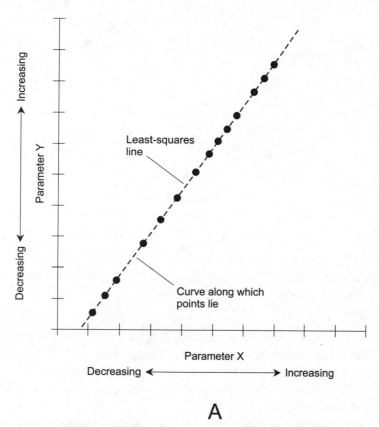

A

Fig. 7-1A. Scatter plot in which the correlation is +1. All the points lie along a smooth curve that happens to be a straight line.

CORRELATION AND OUTLIERS

In some scatter plots, the points are concentrated near smooth curves or lines, although it is rare for any scatter plot to contain points as orderly as

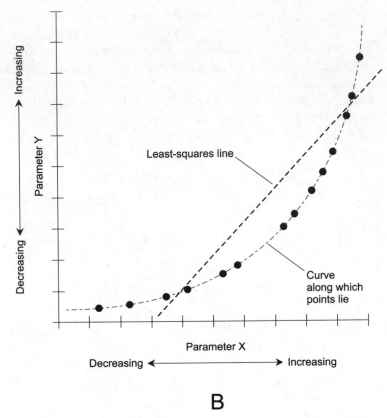

B

Fig. 7-1B.   Scatter plot in which the correlation is not +1. The points lie along a smooth curve, but that curve is not a straight line.

those shown in Fig. 7-1A or B. Once in a while, you'll see a scatter plot in which almost all of the points lie near a straight line, but there are a few points that are far away from the main group. Stray points of this sort are known as *outliers*. These points are, in some ways, like the outliers found in statistical distributions.

One or two "extreme outliers" can greatly affect the correlation between two variables. Consider the example of Fig. 7-2. This is a scatter plot in which all but two of the points are in exactly the same positions as they are in Fig. 7-1A. But the two outliers are both far from the least-squares line. These points happen to be at equal distances (indicated by *d*) from the line, so their net effects on the position of the line cancel each other. Thus, the least-squares line in Fig. 7-2 is in the same position as the least-squares line in Fig. 7-1A. But the correlation values are much different. In Fig. 7-1A, $r =$ +1. In the situation shown by Fig. 7-2, r is much smaller than +1.

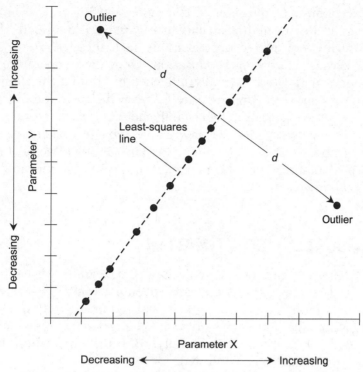

Fig. 7-2.   Scatter plot in which only two outliers drastically reduce the correlation, even though they do not affect the position of the least-squares line. Compare with Fig. 7-1A.

CORRELATION AND DEFINITION OF VARIABLES

Here's another important rule concerning correlation. It doesn't matter which variable is defined as dependent and which variable is defined as independent. If the definitions of the variables are interchanged, and nothing about the actual scenario changes, the correlation remains exactly the same.

Think back to the previous chapter, where we analyzed the correlation between average monthly temperatures and average monthly rainfall amounts for two cities. When we generated the scatter plots, we plotted temperature on the horizontal axis, and considered temperature to be an independent variable. However, we could just as well have plotted the rainfall amounts on the horizontal axis, and defined them as the independent variables. The resulting scatter plots would have looked different, but upon mathematical analysis, the correlation figures would have come out the same.

Sometimes a particular variable lends itself intuitively to the role of the independent variable. (Time is an excellent example of this, although there are some exceptions.) In the cases of Happyton and Blissville from the previous chapter, it doesn't matter much which variable is considered independent and which is considered dependent. In fact, these very labels can be misleading, because they suggest causation. Does the temperature change, over the course of the year, actually influence the rainfall in Happyton or Blissville? If so, the effects are opposite between the two cities. Or is it the other way around – rainfall amounts influence the temperature? Again, if that is true, the effects are opposite between the two cities. There is something a little weird about either assumption. Perhaps another factor, or even a combination of multiple factors, influences both the temperature and the rainfall in both towns.

UNITS (USUALLY) DON'T MATTER

Here's an interesting property of correlation. The units we choose don't matter, as long as they express the same phenomenon or characteristic. If the measurement unit of either variable is changed in size but not in essence, the appearance of a bar graph or scatter plot changes. The plot is "stretched" or "squashed" vertically or horizontally. But the correlation figure, r, between the two variables is unaffected.

Think back again to the last chapter, and the scatter plots of precipitation versus temperature for Happyton and Blissville. The precipitation amounts are indicated in centimeters per month, and the temperatures are shown in degrees Celsius. Suppose the precipitation amounts were expressed in inches per month instead. The graphs would look a little different, but upon analysis by a computer, the correlation figures would turn out the same. Suppose the temperatures were expressed in degrees Fahrenheit. Again, the graphs would look different, but r would not be affected. Even if the average monthly rainfall were plotted in miles per month and the temperatures in degrees Kelvin (where 0 K represents absolute zero, the coldest possible temperature), the value of r would be the same.

We must be careful when applying this rule. The sizes of the units can be changed, but the quantities or phenomena they represent must remain the same. Therefore, if we were to plot the average rainfall in inches, centimeters, or miles per week rather than per month, we could no longer be sure the correlation would remain the same. The scatter plots would no longer show the same functions. The variable on the vertical scale – rainfall averaged over weekly periods rather than over monthly periods – would no longer represent the same thing. This is a subtle distinction, but it makes a critical difference.

PROBLEM 7-1

Suppose the distances of the outliers from the least-squares line from Fig. 7-2 are cut in half (to $d/2$ rather than d), as shown in Fig. 7-3. What effect will this have on the correlation?

SOLUTION 7-1

It will increase the correlation, because the average distances of all the points from the least-squares line will be smaller.

PROBLEM 7-2

Suppose that one of the outliers is removed in the scenario of Fig. 7-3. Will this affect the position of the least-squares line?

SOLUTION 7-2

Yes. If the upper-left outlier is removed, the position of the least-squares line will be displaced slightly downward and to the right; if the lower-right outlier is removed, the least-squares line will be displaced slightly upward and to the left.

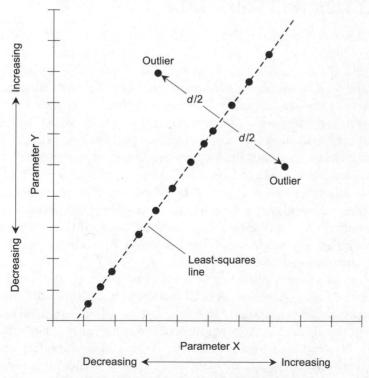

Fig. 7-3. Illustration for Problems 7-1 and 7-2.

Causes and Effects

When two things are correlated, it's tempting to think that there is a cause-and-effect relationship involved. Examples abound; anyone who listens to the radio, reads newspapers, or watches television these days can't escape them. Sometimes the cause–effect relationships aren't directly stated, but only implied. "Take this pill and you'll be happy all the time. Avoid these foods and you won't die of a heart attack."

Some of the cause–effect blurbs we hear every day must sound ridiculous to people who have grown up in cultures radically different from ours. "Drink this fizzy liquid and you'll never get dirt under your fingernails. Eat this creamy white stuff and the hair on your ears will go away." The implication is always the same. "Do something that causes certain people to make a profit, and your life will improve." Sometimes there really is a cause–effect relationship. Sometimes there isn't. Often, we don't know.

CORRELATION AND CAUSATION

Let's boil down a correlation situation to generic terms. That way, we won't be biased (or deluded) into inferring causation. Suppose two phenomena, called X and Y, vary in intensity with time. Figure 7-4 shows a relative graph of the variations in both phenomena. The phenomena change in a manner that is positively correlated. When X increases, so does Y, in general. When Y decreases, so does X, in general.

Is causation involved in the situation shown by Fig. 7-4? Maybe! There are four possible ways that causation can exist. But perhaps there is no cause-and-effect relationship. Maybe Fig. 7-4 shows a coincidence. If there were 1000 points on each plot, there would be a better case for causation. As it is, there are only 12 points on each plot. It is possible these points represent a "freak scenario." There is also a more sinister possibility: The 12 points in each plot of Fig. 7-4 might have been selected by someone with a vested interest in the outcome of the analysis.

When we assign real phenomena or observations to the variables in a graph such as Fig. 7-4, we can get ideas about causation. But these ideas are not necessarily always right. In fact, intense debate often takes place in scientific, political, and even religious circles concerning whether or not a correlation between two things is the result of cause-and-effect, and if so, how the cause-and-effect actually operates. And how do we know that the data itself is not biased?

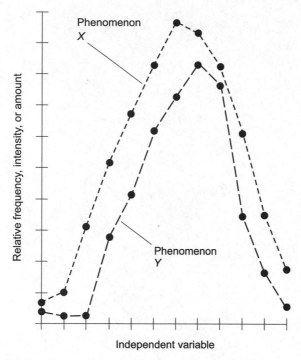

Fig. 7-4.   The two phenomena shown here, *X* and *Y*, are obviously correlated. Does this imply causation?

In the examples that follow, we'll rule out the bias factor and assume that all data has been obtained with the intent of pursuing truth. There are myriad ways in which data can be warped and rigged to distort or cover up truth, but we'll let sociologists, psychologists, and criminologists worry about that.

X causes *Y*

Cause-and-effect relationships can be illustrated using arrows. Figure 7-5A shows the situation where changes in phenomenon *X* directly cause changes in phenomenon *Y*. You can doubtless think of some scenarios. Here's a good real-life example.

Suppose the independent variable, shown on the horizontal axis in Fig. 7-4, is the time of day between sunrise and sunset. Plot *X* shows the relative intensity of sunshine during this time period; plot *Y* shows the relative temperature over that same period of time. We can argue that the brilliance of the sunshine causes the changes in temperature. There is some time lag in the

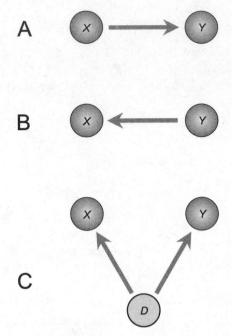

Fig. 7-5. At A, X causes Y. At B, Y causes X. At C, D causes both X and Y.

temperature function; this is to be expected. The hottest part of the day is usually a little later than the time when the sunshine is most direct.

It's harder to believe that there's a cause-and-effect relationship in the other direction. It is silly to suggest that temperature changes cause differences in the brilliance of the sunlight reaching the earth's surface. Isn't it? Maybe, but maybe not. Suppose heating causes the clouds to clear, resulting in more sunlight reaching the surface (Y causes X). Maybe there's something to this sort of argument, but most meteorologists would say that the former relation better represents reality. Bright sunshine heats things up. That's obvious.

Y causes X

Imagine that the horizontal axis in Fig. 7-4 represents 12 different groups of people in a medical research survey. Each hash mark on the horizontal axis represents one group. Plot X is a point-to-point graph of the relative number of fatal strokes in a given year for the people in each of the 12 groups; plot Y is a point-to-point graph of the relative average blood pressure levels of the people in the 12 groups during the same year. (These are hypothetical graphs, not based on any real historical experiments, but a real-life survey might come up with results something like this. Medical research has shown a correlation between blood pressure and the frequency of fatal strokes.)

Is there a cause-and-effect relationship between the value of X and the value of Y here? Most doctors would answer with a qualified yes: variations in Y cause the observed variations in X (Fig. 7-5B). Simply put: high blood pressure can cause fatal strokes, in the sense that, if all other factors are equal, a person with high blood pressure is more likely to have a fatal stroke than a person identical in every other respect, but with normal blood pressure.

What about the reverse argument? Can fatal strokes cause high blood pressure (X causes Y)? No. That's clearly absurd.

Complications

Experts in meteorology and medical science who read this may, at this point, be getting a little nervous. Aren't the above scenarios oversimplified? Yes, they are. The cause-and-effect relationships described aren't "pure." In real life, "pure" cause–effect events, where there is one certain cause and one inevitable effect, rarely occur.

The brightness of the sunshine is not, all by itself, the only cause-and-effect factor in the temperature during the course of a day. A nearby lake or ocean, the wind direction and speed, and the passage of a weather front can all have an effect on the temperature at any given location. We've all seen the weather clear and brighten, along with an abrupt drop in temperature, when a strong front passes by. The sun comes out, and it gets cold. That defies the notion that bright sun causes things to heat up, even though the notion, in its "pure" form where all other factors are equal, is valid. The problem is that other factors are not always equal!

In regards to the blood-pressure-versus-stroke relationship, there are numerous other factors involved, and scientists aren't sure they know them all yet. New discoveries are constantly being made in this field. Examples of other factors that might play cause-and-effect roles in the occurrence of fatal strokes include nutrition, stress, cholesterol level, body fat index, presence or absence of diabetes, age, and heredity. A cause-and-effect relationship (Y causes X) exists, but it is not "pure."

D causes both X and Y

Now suppose that the horizontal axis in Fig. 7-4 represents 12 different groups of people in another medical research survey. Again, each hash mark on the horizontal axis represents one group. Plot X is a point-to-point graph of the relative number of heart attacks in a given year for the people in each of the 12 groups; plot Y is a point-to-point graph of the relative average blood cholesterol levels of the people in the 12 groups during the same year. As in the stroke scenario, these are hypothetical graphs. But they're plausible. Medicine has shown a correlation between blood cholesterol and the frequency of heart attacks.

Before I make enemies in the medical profession, let me say that the purpose of this discussion is not to resolve the cholesterol-versus-heart-disease issue, but to illustrate complex cause–effect relationships. It's easier to understand a discussion about real-life factors than to leave things entirely generic. I do not have the answer to the cholesterol-versus-heart-disease riddle. If I did, I'd be writing a different book.

When scientists first began to examine the hearts of people who died of heart attacks in the early and middle 1900s, they found "lumps" called *plaques* in the arteries. It was theorized that plaques caused the blood flow to slow down, contributing to clots that eventually cut off the blood to part of the heart, causing tissue death. The plaques were found to contain cholesterol. Evidently, cholesterol could accumulate inside the arteries. Consulting data showing a correlation between blood cholesterol levels and heart attacks, scientists got the idea that if the level of cholesterol in the blood could be reduced, the likelihood of a heart attack would also go down. The theory was that fewer or smaller plaques would form, reducing the chances of clot formation that could obstruct an artery. At least, such was the hope.

The obvious first point of attack was to tell heart patients to reduce the amount of cholesterol in their food, hoping that this change in eating behavior would cause blood cholesterol levels to go down. In many cases, a low-cholesterol diet did bring down blood cholesterol levels. (Later, special drugs were developed that had the same effect.) Studies continue along these lines. It is becoming apparent that reducing the amount of cholesterol in the diet, mainly by substituting fruits, vegetables, and whole grains for cholesterol-rich foods, can reduce the levels of cholesterol in the blood. This type of dietary improvement can apparently also reduce the likelihood that a person will have a heart attack in the next year, or two, or three. There's more than mere correlation going on here. There's causation, too. But how much causation is there? And in what directions does it operate?

Let's call the amount of dietary cholesterol "factor D." According to current popular medical theory, there is a cause-and-effect relation between this factor and both X and Y. Some studies have indicated that, all other things being equal, people who eat lots of cholesterol-rich foods have more heart attacks than people whose diets are cholesterol-lean. The scenario is shown in Fig. 7-5C. There is a cause-and-effect relation between factor D (the amount of cholesterol in the diet) and factor X (the number of heart attacks); there is also a cause-and-effect relation between factor D and factor Y (the average blood cholesterol level). But most scientists would agree that it's an oversimplification to say this represents the whole picture. If you become a strict vegetarian and avoid cholesterol-containing foods altogether, there's no guarantee that you'll never have a heart attack. If you eat steak and eggs every day for breakfast, it doesn't mean that you are doomed to have a heart attack. The cause-and-effect relationship exists, but it's not "pure," and it's not absolute.

Multiple factors cause both X and Y

If you watch television shows where the advertising is aimed at middle-aged and older folks, you'll hear all about cholesterol and heart disease – probably a good deal more than you want to hear. High cholesterol, low cholesterol, HDL, LDL, big particles, small particles. You might start wondering whether you should go to a chemistry lab rather than a kitchen to prepare your food. The cause-and-effect relationship between cholesterol and heart disease is complicated. The more we learn, it appears, the less we know.

Let's introduce and identify three new friends: factors S, H, and E. Factor S is named "Stress" (in the sense of anxiety and frustration), factor H is named "Heredity" (in the sense of genetic background), and factor E is named "Exercise" (in the sense of physical activity). Over the past several decades, cause-and-effect relationships have been suggested between each of these factors and blood cholesterol, and between each of these factors and the frequency of heart attacks. Figure 7-6 illustrates this sort of "cause-and-effect web." Proving the validity of each link – for example, whether or not stress, all by itself, can influence cholesterol in the blood – is a task for future researchers. But all of the links shown in the diagram have been suggested by somebody.

COINCIDENCE

The existence of correlation between two phenomena doesn't necessarily imply any particular cause-and-effect scenario. Two phenomena can be correlated because of a sheer coincidence. This is most likely to happen when the amount of data obtained is not adequate. In the case of blood cholesterol versus the frequency of heart attacks, test populations have traditionally contained thousands of elements (people). The researchers are justified in

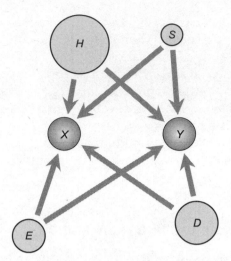

Fig. 7-6.   A scenario in which there are multiple causes *D*, *S*, *H*, and *E* for both *X* and *Y*, not necessarily all of equal effect or importance.

their conclusions that such correlation is not the product of coincidence. It is reasonable to suppose that some cause-and-effect interaction exists. Researchers are still busy figuring out exactly how it all works, and if they ever get it completely unraveled, it's a good bet that an illustration of the "cause-and-effect web" will look a lot more complicated than Fig. 7-6.

Correlation indicates that things take place more or less in concert with one another. This allows us to predict certain future events with varying degrees of accuracy. But there is another form of order that can be found in nature. This form of order, defined by a new science called *chaos theory*, illustrates that some phenomena, totally unpredictable and which defy statistical analysis in the short term or on a small scale, can nevertheless be highly ordered and predictable on a large scale. In a moment, we'll look at this.

PROBLEM 7-3
What are some reasonable cause-and-effect relationships that might exist in Fig. 7-6, other than those shown or those directly between *X* and *Y*? Use arrows to show cause and effect, and use the abbreviations shown.

SOLUTION 7-3
Consider the following. Think about how you might conduct statistical experiments to check the validity of these notions, and to determine the extent of the correlation.

- H → S (Hypothesis: Some people are born more stress-prone than others.)
- H → D (Hypothesis: People of different genetic backgrounds have developed cultures where the diets are dramatically different.)
- E → S (Hypothesis: Exercise can relieve or reduce stress.)
- E → D (Hypothesis: Extreme physical activity makes people eat more food, particularly carbohydrates, because they need more.)
- D → S (Hypothesis: Bad nutritional habits can worsen stress. Consider a hard-working person who lives on coffee and potato chips, versus a hard-working person who eats mainly fish, fruits, vegetables, and whole grains.)
- D → E (Hypothesis: People with good nutritional habits get more exercise than people with bad nutritional habits.)

PROBLEM 7-4
What are some cause-and-effect relationships in the diagram of Fig. 7-6 that are questionable or absurd?

SOLUTION 7-4
Consider the following. Think about how you might conduct statistical experiments to find out whether or not the first three of these might be moved into the preceding category.

- H → E (Question: Do people of certain genetic backgrounds naturally get more physical exercise than people of other genetic backgrounds?)
- S → E (Question: Can stress motivate some types of people to exercise more, yet motivate others to exercise less?)
- S → D (Question: Do certain types of people eat more under stress, while others eat less?)
- S → H (Obviously absurd. Stress can't affect a person's heredity!)
- E → H (Obviously absurd. Exercise can't affect a person's heredity!)
- D → H (Obviously absurd. Dietary habits can't affect a person's heredity!)

Chaos, Bounds, and Randomness

Have you ever noticed that events seem to occur in bunches? This is more than your imagination. A few decades ago, this phenomenon was analyzed by Benoit Mandelbrot, an engineer and mathematician who worked for

International Business Machines (IBM). Mandelbrot noticed that similar patterns are often found in apparently unrelated phenomena, such as the fluctuations of cotton prices and the distribution of personal incomes. His work gave birth to the science of *chaos theory*.

WAS ANDREW "DUE"?

In the early summer of 1992, south Florida hadn't had a severe hurricane since Betsy in 1965. The area around Miami gets a minimal hurricane once every 7 or 8 years on the average, and an extreme storm once or twice a century. Was Miami "due" for a hurricane in the early 1990s? Was it "about time" for a big blow? Some people said so. By now you should know enough about probability to realize that 1992 was no more or less special, in that respect, than any other year. In fact, as the hurricane season began in June of that year, the experts predicted a season of below-normal activity.

The so-called "law of averages" (which is the basis for a great deal of misinformation and deception) seemed to get its justice on August 24, 1992. Hurricane Andrew tore across the southern suburbs of Miami and the Everglades like a cosmic weed-whacker, and became the costliest hurricane ever to hit the United States up to that date. Did the severity of Andrew have anything to do with the lack of hurricanes during the previous two and a half decades? No. Did Andrew's passage make a similar event in 1993 or 1994 less likely than it would have been if Andrew had not hit south Florida? No. There could have been another storm like Andrew in 1993, and two more in 1994. Theoretically, there could have been a half dozen more like it later in 1992!

Have you ever heard about a tornado hitting some town, followed three days later by another one in the same region, and four days later by another, and a week later by still another? Have you ever flipped a coin for a few minutes and had it come up "heads" 18 times in a row, even though you'd normally have to flip it for days to expect such a thing to happen? Have you witnessed some vivid example of "event-bunching," and wondered if anyone will ever come up with a mathematical theorem that tells us why this sort of thing seems to happen so often?

SLUMPS AND SPURTS

Athletes such as competitive swimmers and runners know that improvement characteristically comes in spurts, not smoothly with the passage of time. An example is shown in Fig. 7-7 as a graph of the date (by months during a

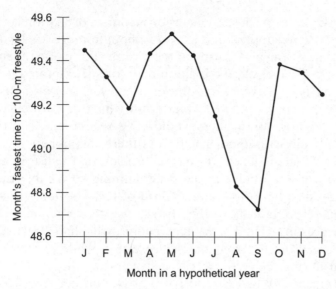

Fig. 7-7. Monthly best times (in seconds) for a swimmer whose specialty is the 100-m freestyle, plotted by month for a hypothetical year. Illustration for Problems 7-5 and 7-6.

hypothetical year) versus time (in seconds) for a hypothetical athlete's 100-meter (100-m) freestyle swim. The horizontal scale shows the month, and the vertical scale shows the swimmer's fastest time in that month.

Note that the swimmer's performance does not improve for a while, and then suddenly it does. In this example, almost all of the improvement occurs during the summer training season. That's not surprising, but another swimmer might exhibit performance that worsens during the same training season. Does this irregularity mean that all the training done during times of flat performance is time wasted? The coach will say no! Why does improvement take place in sudden bursts, and not gradually with time? Sports experts will tell you they don't know. Similar effects are observed in the growth of plants and children, in the performance of corporate sales departments, and in the frequency with which people get sick. This is "just the way things are."

CORRELATION, COINCIDENCE, OR CHAOS?

Sometime in the middle of the 20th century, a researcher noticed a strong correlation between the sales of television sets and the incidence of heart attacks in Great Britain. The two curves followed remarkably similar contours. In fact the shapes of the graphs were, peak-for-peak and valley-for-valley, almost identical.

It is tempting to draw hasty conclusions from a correlation such as this. It seems reasonable to suppose that as people bought more television sets, they spent more time sitting and staring at the screens; this caused them to get less exercise; the people's physical condition therefore deteriorated; this rendered them more likely to have heart attacks. But even this argument, if valid, couldn't explain the uncanny exactness with which the two curves followed each other, year after year. There would have been a lag effect if television-watching really did cause poor health, but there was none.

Do television sets emit electromagnetic fields that cause immediate susceptibility to a heart attack? Is the programming so terrible that it causes immediate physical harm to viewers? Both of these notions seem "far-out." Were the curves obtained by the British researcher coincident for some unsuspected reason? Could it be that people who had heart attacks were told by their doctors to avoid physical exertion while recovering, and this caused them to buy television sets to help pass the time? Or was the whole thing a coincidence? Was there no true correlation between television sales and heart attacks, a fact that would have become apparent if the experiment had continued for decades longer or had involved more people?

Now consider this if you dare: Could the correlation between television sales and heart attacks have taken place as a result of some unfathomable cosmic consonance, even in the absence of a cause-and-effect relationship?

Do scientists sometimes search for nonexistent cause-and-effect explanations, getting more and more puzzled and frustrated as the statistical data keeps pouring in, demonstrating the existence of a correlation but giving no clue as to what is responsible for it? Applied to economic and social theory, this sort of correlation-without-causation phenomenon can lead to some scary propositions. Is another world war, economic disaster, or disease pandemic inevitable because that's "just the way things are"? Chaos theory suggests that the answer to some of these questions is yes!

SCALE-RECURRENT PATTERNS

Benoit Mandelbrot noticed that patterns tend to recur over various time scales. Large-scale and long-range changes take place in patterns similar to those of small-scale and short-term changes. Events occur in bunches; the bunches themselves take place in similar bunches following similar patterns. This effect exists both in the increasing scale and in the decreasing scale.

Have you noticed that high, cirrostratus clouds in the sky resemble the clouds in a room where someone has recently lit up a cigar? Or that these clouds look eerily like the interstellar gas-and-dust clouds that make up diffuse nebulae in space? Patterns in nature often fit inside each other as if they were nested geometric shapes, as if the repetition of patterns over scale takes place because of some principle ingrained in nature itself. This is evident when you look at the so-called *Mandelbrot set* (Fig. 7-8) using any of the numerous zooming programs available on the Internet. This set arises from a simple

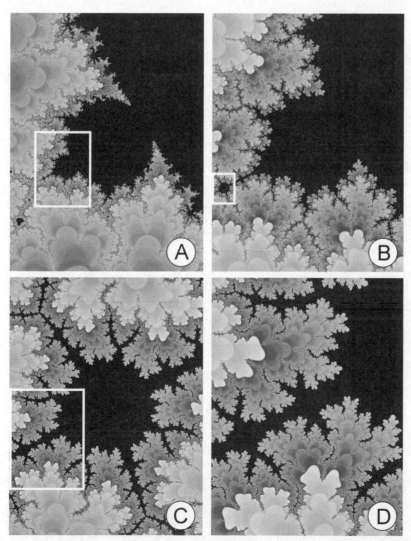

Fig. 7-8. Portions of the Mandelbrot set, viewed with repeated magnification (views enlarge progressively from A through D). Numerous free programs for generating images of this sort are available on the Internet. These images were created using Fractint for Windows.

mathematical formula, yet it is infinitely complicated. No matter how much it is magnified – that is, however closely we zoom in on it – new patterns appear. There is no end to it! Yet the patterns show similarity at all scales.

The images in Fig. 7-8 were generated with a freeware program called *Fractint*. This program was created by a group of experts called the Stone Soup Team. The program itself is copyrighted, but images created by any user become the property of that user.

PROBLEM 7-5

Does the tendency of athletic performance to occur in "spurts" mean that a gradual improvement can never take place in any real-life situation? For example, is it impossible that the curve for the swimmer's times (Fig. 7-7) could look like either the solid or dashed lines in Fig. 7-9 instead?

SOLUTION 7-5

Gradual, smooth improvement in athletic performance is possible. Either of the graphs (the straight, dashed line or the solid, smooth curve) in Fig. 7-9 could represent a real-life situation. But orderly states of affairs such as this are less common than more chaotic types of variations such as that shown in Fig. 7-7.

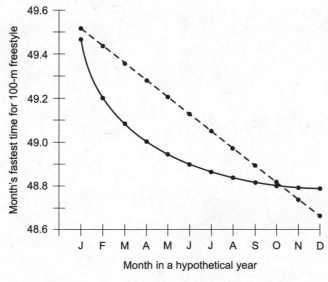

Fig. 7-9.   Illustration for Problems 7-5 and 7-6.

PROBLEM 7-6

What is the fastest possible time for the 100-m freestyle swimming event, as implied by the graphs in Fig. 7-7 or Fig. 7-9?

SOLUTION 7-6

Neither of these graphs logically implies that there is a specific time that represents the fastest possible 100-m swim. It can be mathematically proven that there exists a *maximum unswimmable time* for this event. But determining the actual time quantitatively, and then proving that the determination is valid, is another problem.

THE MAXIMUM UNSWIMMABLE TIME

If our hypothetical swimmer keeps training, how fast will he eventually swim the 100-m freestyle? We already know that he can do it in a little more than 48 seconds. What about 47 seconds? Or 46 seconds? Or 45 seconds? There are obvious *lower bounds* to the time in which the 100-m freestyle can be swum by a human. It's a good bet that no one will ever do it in 10 seconds. How about 11 seconds? Or 12? Or 13? How about 20 seconds? Or 25? Or 30? If we start at some ridiculous figure such as 10 seconds and keep increasing the number gradually, we will at some point reach a figure – let's suppose for the sake of argument that it is 41 seconds – representing the largest whole number of seconds too fast for anyone to swim the 100-m freestyle.

Once we have two whole numbers, one representing a swimmable time (say 42 seconds) and the next smaller one representing an unswimmable time (say 41 seconds), we can refine the process down to the tenth of a second, and then to the hundredth, and so on indefinitely. There is some time, exact to however small a fraction of a second we care to measure it, that represents the maximum unswimmable time (MUST) that a human being can attain for the 100-m freestyle swim. Figure 7-10 shows an educated estimate (translation: wild guess) for this situation.

No one knows the exact MUST for the 100-m freestyle, and a good argument can be made for the assertion that we *cannot* precisely determine it. But such a time nevertheless exists. How do we know that there is a MUST for the 100-m freestyle, or for any other event in any other timed sport? A well-known theorem of mathematics, called the *theorem of the greatest lower bound*, makes it plain: "If there exists a lower bound for a set, then there exists a *greatest lower bound* (GLB) for that set." A more technical term for GLB is *infimum*. In this case, the set in question is the set of "swimmable times" for the 100-m freestyle. The lower bounds are the "unswimmable times."

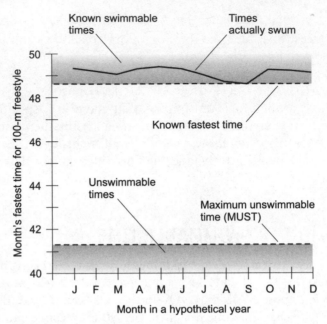

Fig. 7-10.   According to the theorem of the greatest lower bound, there exists a maximum
unswimmable time (MUST) for the 100-meter freestyle. (The time shown in this
graph is an educated guess.)

What's the probability that a human being will come to within a given
number of seconds of the MUST for the 100-m freestyle in, say, the next 10
years, or 20 years, or 50 years? Sports writers may speculate on it; physicians
may come up with ideas; swimmers and coaches doubtless have notions too.
But anyone who makes a claim in this respect is only guessing. We can't say
"The probability is 50% that someone will swim the 100-m freestyle in so-
and-so seconds by the year such-and-such." Remember the old probability
fallacy from Chapter 3! For any theoretically attainable time, say 43.50
seconds, one of two things will happen: someone will swim the 100-m freestyle that fast someday, or else no one will.

THE BUTTERFLY EFFECT

The tendency for small events to have dramatic long-term and large-scale
consequences is called the *butterfly effect*. It gets its name from a hypothetical
question that goes something like this: Can a butterfly taking off in China
affect the development, intensity, and course of a hurricane 6 months later in
Florida? At first, such a question seems ridiculous. But suppose the butterfly

creates a tiny air disturbance that produces a slightly larger one, and so on, and so on, and so on. According to butterfly-effect believers, the insect's momentary behavior could be the trigger that ultimately makes the difference between a tropical wave and a killer cyclone.

We can never know all the consequences of any particular event. History happens once and only once. We can't make repeated trips back in time and let fate unravel itself multiple times, after tweaking this or that little detail. But events can conspire, or have causative effects over time and space, in such a manner as to magnify the significance of tiny events in some circumstances. There are computer models to show it.

Suppose you go out biking in the rain and subsequently catch a cold. The cold develops into pneumonia, and you barely survive. Might things have turned out differently if the temperature had been a little warmer, or if it had rained a little less, or if you had stayed out for a little less time? There is no practical way to tell which of these tiny factors are critical and which are not. But computer models can be set up, and programs run, that in effect "replay history" with various parameters adjusted. In some cases, certain variables have threshold points where a tiny change will dramatically affect the distant future.

SCALE PARALLELS

In models of chaos, patterns are repeated in large and small sizes for an astonishing variety of phenomena. A good example is the comparison of a spiral galaxy with a hurricane. The galaxy's stars are to the hurricane's water droplets as the galaxy's spiral arms are to the hurricane's rainbands. The eye of the hurricane is calm and has low pressure; everything rushes in towards it. The water droplets, carried by winds, spiral inward more and more rapidly as they approach the edge of the eye. In a spiral galaxy, the stars move faster and faster as they fall inward toward the center. A satellite photograph of a hurricane, compared with a photograph of a spiral galaxy viewed face-on, shows similarities in the appearance of these systems.

Air pressure and gravitation can both, operating over time and space on a large scale, produce the same kind of spiral. Similar spirals can be seen in the Mandelbrot set and in other similar mathematically derived patterns. The *Spiral of Archimedes* (a standard spiral easily definable in analytic geometry) occurs often in nature, and in widely differing scenarios. It's tempting to believe that these structural parallels are more than coincidences, that there is a cause-and-effect relationship. But what cause-and-effect factor can make a spiral galaxy in outer space look and revolve so much like a hurricane on the surface of the earth?

PROBLEM 7-7
Can the MUST scenario, in which there is a greatest lower bound, apply in a reverse sense? Can there be, for example, a *minimum unattainable temperature* (MUTT) for the planet earth?

SOLUTION 7-7
Yes. The highest recorded temperature on earth, as of this writing, is approximately 58°C (136°F). Given current climatic conditions, it's easy to imagine an unattainable temperature, for example, 500°C. We then start working our way down from this figure. Clearly, 400°C is unattainable, as is 300°C, and also 200°C (assuming runaway global warming doesn't take place, in which our planet ends up with an atmosphere like that of Venus). What about 80°C? What about 75°C? A theorem of mathematics, called the *theorem of the least upper bound*, makes it plain: "If there exists an upper bound for a set, then there exists a *least upper bound* (LUB) for that set." This means there is a MUTT for our planet, given current climatic conditions. Figuring out an exact number, in degrees, for the MUTT is another problem.

THE MALTHUSIAN MODEL

Chaos theory has been applied in grim fashion to describe the characteristics of the earth's population growth. Suppose we want to find a function that can describe world population versus time. The simplest model allows for an exponential increase in population, but this so-called *Malthusian model* (named after its inventor, Thomas Malthus) does not incorporate factors such as disease pandemics, world wars, or the collision of an asteroid with the planet.

The Malthusian model is based on the idea that the world's human population increases geometrically in the way bacteria multiply, while the world's available supply of food and other resources increase arithmetically. It is easy to see that a pure Malthusian population increase can only go on for a certain length of time. When a certain critical point is reached, the population will no longer increase, because the earth will get too crowded and there won't be enough resources to keep people alive. What will happen then? Will the population level off smoothly? Will it decline suddenly and then increase again? Will it decline gradually and then stay low? The outcome depends on the values we assign to certain parameters in the function we ultimately find that describes population versus time.

A BUMPY RIDE

The limiting process for any population-versus-time function depends on the extent of the disparity between population growth and resource growth. If we consider the earth's resources to be finite, then the shape of the population-versus-time curve depends on how fast people reproduce until a catastrophe occurs. As the reproduction rate goes up – as the "function is driven harder" – the time period until the first crisis decreases, and the ensuing fluctuations become more and more wild. In the worst cases, the predictions become dire indeed.

The Malthusian equation for population increase is:

$$x_{n+1} = rx_n (1 - x_n)$$

where n is a whole number starting with $n = 0$, and r is a factor that represents the rate of population increase. (This is not the same r factor that represents correlation, defined earlier in this chapter.) Statisticians, social scientists, biologists, mathematicians, and even some politicians have run this formula through computers for various values of r, in an attempt to predict what would happen to the world's population as a function of time on the basis of various degrees of "population growth pressure." It turns out that a leveling-off condition occurs when the value of r is less than about 2.5. The situation becomes more complicated and grotesque with higher values of r. As the value of the r factor increases, the function is "driven harder," and the population increases with greater rapidity – until a certain point in time. Then chaos breaks loose.

According to computer models, when the r factor is low, the world population increases, reaches a peak, and then falls back. Then the population increases again, reaches another peak, and undergoes another decline. This takes place over and over but with gradually diminishing wildness. Thus, a *damped oscillation* occurs in the population function as it settles to a steady state (Fig. 7-11A).

In real life, the r factor can be kept low by strict population control and public education. Conversely, the r factor could become higher if all efforts at population control were abandoned. Computers tell us with unblinking screens what they "think" will happen then. If the value of r is large enough, the ultimate world population does not settle down, but oscillates indefinitely between limiting values. The amplitude and frequency of the oscillation

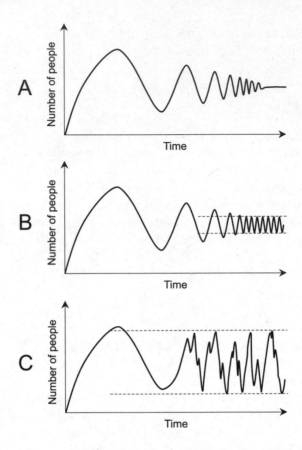

Fig. 7-11. Number of people in the world as a function of time. At A, a small *r* factor produces eventual stabilization. At B, a large *r* factor produces oscillation within limits. At C, a very large *r* factor produces chaotic fluctuation within limits.

depends on how large the *r* factor is allowed to get (Fig. 7-11B). At a certain critical value for the *r* factor, even this vestige of orderliness is lost, and the population-versus-time function fluctuates crazily, never settling into any apparent oscillation frequency, although there are apparent maximum and minimum limits to the peaks and valleys (Fig. 7-11C).

A graph in which the world's ultimate human population is plotted on the vertical (dependent-variable) axis and the *r* factor is plotted on the horizontal (independent-variable) axis produces a characteristic pattern something like the one shown in Fig. 7-12. The function breaks into oscillation when the *r* factor reaches a certain value. At first this oscillation has defined frequency and amplitude. But as *r* continues to increase, a point is reached where the oscillation turns into noise. As an analogy, think about what happens when

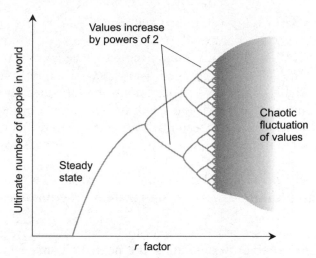

Fig. 7-12. Generalized, hypothetical graph showing "final" world population as a function of the relative r factor.

the audio gain of a public-address system is increased until feedback from the speakers finds its way to the microphone, and the speakers begin to howl. If the audio gain is turned up higher, the oscillations get louder. If the system is driven harder still, the oscillations increase in fury until, in the absence of all restraint, the system roars like thunder.

Does the final population figure in the right-hand part of Fig. 7-12 truly represent unpredictable variation between extremes? If the computer models are to be believed, it does. By all indications, the gray area in the right-hand part of Fig. 7-12 represents a sort of *randomness*.

WHAT IS RANDOMNESS?

In statistical analysis, there is often a need to obtain sequences of values that occur at random. What constitutes randomness? Here's one definition that can be applied to single-digit numbers:

- A sequence of digits from the set {0, 1, 2, 3, 4, 5, 6, 7, 8, 9} can be considered random if and only if, given any digit in the sequence, there exists no way to predict the next one.

At first thought, the task of generating a sequence of random numbers in this way seems easy. Suppose we chatter away, carelessly uttering digits from 0 to 9. Everyone has a leaning or preference for certain digits or sequences of digits, such as 5 or 58 or 289 or 8827. If a sequence of digits is truly random, then over long periods a given digit x will occur exactly 10% of the time, a given sequence xy will occur exactly 1% of the time, a given sequence xyz will occur exactly 0.1% of the time, and a given sequence $wxyz$ will occur exactly 0.01% of the time. These percentages, over time, should hold for all possible sequences of digits of the given sizes, and similar rules should hold for sequences of any length. But if you speak or write down or keypunch digits for a few days and record the result, it's a good bet that this will not be the case. (Your imagination may seem wild to you, but there is some order to it no matter what.)

Here's another definition of randomness. This definition is based on the idea that all artificial processes contain inherent orderliness:

- In order for a sequence of digits to be random, there must exist no algorithm capable of generating the next digit in a sequence, on the basis of the digits already generated in that sequence.

According to this definition, if we can show that any digit in a sequence is a function of those before it, the sequence is not random. This rules out many sequences that seem random to the casual observer. For example, we can generate the value of the square root of 2 (or $2^{1/2}$) with an algorithm called *extraction of the square root*. This algorithm can be applied to any whole number that is not a perfect square. If we have the patience, and if we know the first n digits of a square root, we can find the $(n + 1)$st digit by means of this process. It works every time, and the result is the same every time. The sequence of digits in the decimal expansion of the square root of 2, as well as the decimal expansions of π, e, or any other irrational number, is the same every time a computer grinds it out. The decimal expansions of irrational numbers are therefore not random-digit sequences.

If the digits in any given irrational number fail to occur in a truly random sequence, where can we find digits that do occur randomly? Is there any such thing? If a random sequence of digits cannot be generated by an algorithm, does this rule out any thought process that allows us to identify the digits? Are we looking for something so elusive that, when we think we've found it, the very fact that we have gone through a thought process to find it proves that we have not? If that is true, how is the statistician to get hold of a random sequence that can actually be used?

In the interest of practicality, statisticians often settle for *pseudorandom* digits or numbers. The prefix *pseudo-* in this context means "pretend" or "for all practical purposes." Computer algorithms exist that can be used to generate strings of digits or numbers that can be considered random in most real-world applications.

THE NET TO THE RESCUE

You can search the Internet and find sites with information about pseudo-random and random numbers. There's plenty of good reading on the Web (as well as plenty of nonsense), and even some downloadable programs that can turn a home computer into a generator of pseudorandom digits. For a good start, go to the Google search engine at *www.google.com*, bring up the page to conduct an advanced search, and then enter the phrase "random number generator." Be careful what you download! Make sure your anti-virus program is effective and up to date. If you are uneasy about downloading stuff from the Web, then don't do it.

A safer way to get random digits is a site maintained by the well-known and respected mathematician Dr. Mads Haahr of Trinity College in Dublin, Ireland. It can be brought up by pointing your Web browser to *www.random.org*. The author describes the difference between pseudorandom and truly random numbers. He also provides plenty of interesting reading on the subject, and links to sites for further research.

Dr. Haahr's Web site makes use of electromagnetic noise to obtain real-time random-number and pseudorandom-number sequences. For this scheme to work, there must exist no sources of orderly noise near enough to be picked up by the receiver. Orderly noise sources include internal combustion engines and certain types of electrical appliances such as old light dimmers. The hissing and crackling that you hear in a radio receiver when it is tuned to a vacant channel is mostly electromagnetic noise from the earth's atmosphere and from outer space. Some electrical noise also comes from the internal circuitry of the radio.

PROBLEM 7-8
Name two poor sources of pseudorandom digits, and two good sources, that can be obtained by practical means.

SOLUTION 7-8
The digits of a known irrational number such as the square root of 20 represent an example of a poor source of pseudorandom digits. These digits are predestined (that is, they already exist and are the same every time they are generated). In addition, they can be produced by a simple machine algorithm.

The repeated spinning of a wheel or rotatable pointer, calibrated in digits from 0 to 9 around the circumference, is another example of a poor way to get a pseudorandom sequence. The wheel can never be spun forcefully enough to randomize the final result. In addition, friction is not likely to be uniform at all points in the wheel bearing's rotation, causing it to favor certain digits over others.

A good source of pseudorandom digits can be obtained by building a special die that has the shape of a *regular dodecahedron*, or geometric solid with 12 identical faces. The faces can be numbered from 0 to 9, with two faces left blank. This die can be repeatedly thrown and the straight-upward-facing side noted. (If that side is blank, the toss is ignored.) The results over time can be tallied as pseudorandom digits. The die must have uniform density throughout, to ensure that it doesn't favor some digits over others.

Another good source of pseudorandom digits is a machine similar to those used in some states' daily televised lottery drawings. A set of 10 lightweight balls, numbered 0 through 9, are blown around by powerful air jets inside a large jar. To choose a digit, the lid of the jar is opened just long enough to let one of the balls fly out. After the digit on the "snagged" ball is noted, the ball is put back into the jar, and the 10 balls are allowed to fly around for a minute or two before the next digit is "snagged."

Quiz

Refer to the text in this chapter if necessary. A good score is 8 correct. Answers are in the back of the book.

1. Refer to the correlation plot of Fig. 7-13. Suppose the dashed line represents the least-squares line for all the solid black points. If a new value is added in the location shown by the gray point *P*, but no other new values are added, what will happen to the least-squares line?
   (a) It will move to the left of the position shown.
   (b) It will move to the right of the position shown.
   (c) Its position will not change from that shown.
   (d) More information is necessary to answer this question.

2. Refer to the correlation plot of Fig. 7-13. Suppose the dashed line represents the least-squares line for all the solid black points. If a new value is added in the location shown by the gray point *Q*, but no other new values are added, what will happen to the least-squares line?
   (a) It will move to the left of the position shown.
   (b) It will move to the right of the position shown.
   (c) Its position will not change from that shown.
   (d) More information is necessary to answer this question.

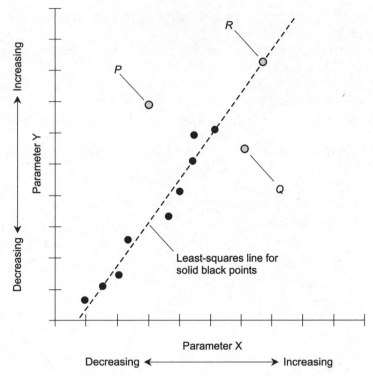

Fig. 7-13. Illustration for Quiz Questions 1 through 3.

3. Refer to the correlation plot of Fig. 7-13. Suppose the dashed line represents the least-squares line for all the solid black points. If a new value is added in the location shown by the gray point R, but no other new values are added, what will happen to the least-squares line?
   (a) It will move to the left of the position shown.
   (b) It will move to the right of the position shown.
   (c) Its position will not change from that shown.
   (d) More information is necessary to answer this question.

4. Theoretically, the earth's population function can level off at a stable value
   (a) if we wait long enough
   (b) if the Malthusian r factor exceeds a certain critical value
   (c) if the Malthusian r factor is kept low enough
   (d) under no circumstances

5. If the correlation between two phenomena X and Y is given by $r = 0$, then an increase in the frequency, intensity, or amount of X

(a) is unrelated to any change in the frequency, intensity, or amount of Y

(b) is usually attended by an increase in the frequency, intensity, or amount of Y

(c) is usually attended by a decrease in the frequency, intensity, or amount of Y

(d) causes a change in the frequency, intensity, or amount of Y

6. According to the Malthusian model, the earth's population, in the absence of limiting factors, would
(a) increase geometrically
(b) increase up to a certain point and then level off
(c) increase linearly
(d) decrease to zero

7. In the plot of Fig. 7-14, the correlation between phenomenon X and phenomenon Y appears to be
(a) positive

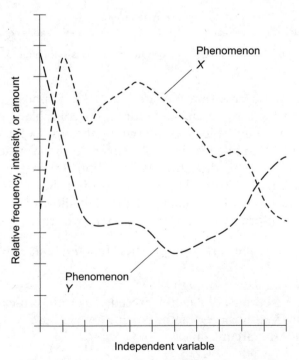

Fig. 7-14. Illustration for Quiz Questions 7 and 8.

(b) negative
(c) chaotic
(d) geometric

8. With respect to the plot shown by Fig. 7-14, which of the following scenarios (a), (b), or (c) is plausible?
    (a) Changes in the frequency, intensity, or amount of X cause changes in the frequency, intensity, or amount of Y.
    (b) Changes in the frequency, intensity, or amount of Y cause changes in the frequency, intensity, or amount of X.
    (c) Changes in the frequency, intensity, or amount of some third factor, Z, cause changes in the frequencies, intensities, and amounts of both X and Y.
    (d) Any of the above scenarios (a), (b), or (c) is plausible.

9. Correlation is a measure of the extent to which the points in a scatter plot
    (a) tend to lie near the origin
    (b) tend to be arranged in a circle
    (c) tend to be arranged along a straight line
    (d) tend to lie near the center of the graph

10. Why are the digits in the decimal expansion of the square root of 10 not truly random?
    (a) Because they can be generated by an algorithm.
    (b) Because the decimal expansion of the square root of 10 cannot be defined.
    (c) Because the decimal expansion of the square root of 10 is a rational number.
    (d) The premise is wrong! They are truly random.

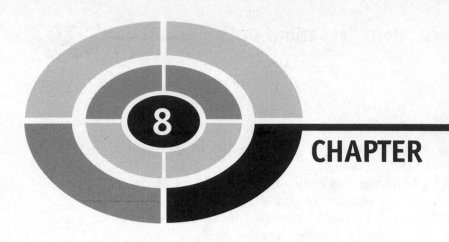

CHAPTER

Some Practical Problems

In this chapter, we'll statistically analyze some hypothetical, but realistic, scenarios. The scope of this chapter is limited to material covered in the preceding chapters.

Frequency Distributions

Imagine that a large class of students is given a quiz. We examine the results in the form of tables and graphs.

PROBLEM 8-1

In our hypothetical class, 130 students take a 10-question quiz. The results are described to us in long-winded verbal form. Here's what we're told: "Nobody missed all the questions (that is, got a score of 0 correct); 4 people got 1 correct answer; 7 people got 2 correct answers; 10 people got 3 correct answers; 15 people got 4 correct answers; 24 people got 5 correct answers; 22 people got 6 correct answers; 24 people got 7 correct answers; 15 people got 8 correct answers; 7 people got 9 correct answers; and 2 people wrote perfect papers (that is, got 10 correct answers)."

Portray these results in the form of a table, showing the test scores in ascending order from top to bottom in the left-hand column, and the absolute frequencies for each score in the right-hand column.

SOLUTION 8-1

Table 8-1 shows the results of the quiz in tabular form. Note that in this depiction, the lowest score is at the top, and the highest score is at the bottom. The table is arranged this way because that's how we are asked to do it.

Table 8-1   Table for Problem 8-1. The lowest score is at the top and the highest score is at the bottom.

Test score	Absolute frequency
0	0
1	4
2	7
3	10
4	15
5	24
6	22
7	24
8	15
9	7
10	2

PROBLEM 8-2
How else can the data from Problem 8-1 be arranged in a table?

SOLUTION 8-2
The quiz results can be portrayed in a table upside-down relative to Fig. 8-1, that is, with the highest score at the top and the lowest score at the bottom (Table 8-2), and it shows us the information just as well. The table can also be arranged with the columns and rows interchanged, so it has 2 rows and 11 columns (not counting the column with the headers). This can be done in either of two ways: the lowest score at the left and the highest score at the right (Table 8-3A), or the highest score at the left and the lowest score at the right (Table 8-3B).

PROBLEM 8-3
Render the data from Problem 8-1 in the form of a vertical bar graph, showing the lowest score at the left and the highest score at the right. Do not put numbers for the absolute frequency values at the tops of the bars.

SOLUTION 8-3
Figure 8-1 shows the results of the quiz as a vertical bar graph, without absolute frequency values shown at the tops of the bars. The advantage of showing the numbers, if there's room to do so, is the fact that it eliminates the

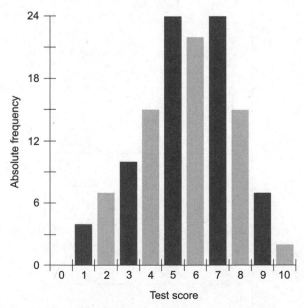

Fig. 8-1.   Illustration for Problems 8-1 and 8-3.

Table 8-2   Table for Problem 8-2.
This is the same data as that shown in
Table 8-1, but with the highest score at
the top and the lowest score at the
bottom.

Test score	Absolute frequency
10	2
9	7
8	15
7	24
6	22
5	24
4	15
3	10
2	7
1	4
0	0

need for the observer having to guess at the values. In this graph, it would be
a "tight squeeze" to show the numbers, and the result would look crowded
and might even cause confusion in reading the graph.

PROBLEM 8-4
Render the data from Problem 8-1 in the form of a horizontal bar graph,
showing the highest score at the top and the lowest score at the bottom.
Include the absolute frequency values at the right-hand ends of the bars.

SOLUTION 8-4
Figure 8-2 shows the results of the quiz as a horizontal bar graph, with the
absolute frequency values indicated at the right-hand ends of the bars. In this

Table 8-3A.   Table for Problem 8-2. This is the same data as that shown in Table 8-1, but with the data arranged horizontally. The lowest score is at the left and the highest score is at the right.

Test score	0	1	2	3	4	5	6	7	8	9	10
Absolute frequency	0	4	7	10	15	24	22	24	15	7	2

Table 8-3B   Another table for Problem 8-2. This is the same data as that shown in Table 8-1, but with the data arranged horizontally. The highest score is at the left and the lowest score is at the right.

Test score	10	9	8	7	6	5	4	3	2	1	0
Absolute frequency	2	7	15	24	22	24	15	10	7	4	0

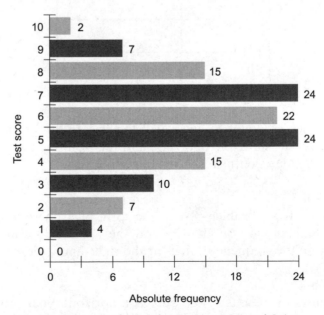

Fig. 8-2.   Illustration for Problems 8-1 and 8-4.

graph, it is not quite so messy to show the numbers, and they provide useful information.

PROBLEM 8-5
Render the data from Problem 8-1 in the form of a point-to-point graph, showing the lowest score at the left and the highest score at the right on the horizontal scale, and showing the absolute frequency referenced to the vertical scale with the lowest values at the bottom and the highest values at the top.

SOLUTION 8-5
Figure 8-3 is an example of such a graph. Data values are shown by the points. The straight lines create an impression of the general shape of the distribution, but are not part of the graph itself.

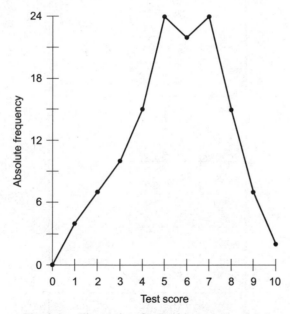

Fig. 8-3.   Illustration for Problems 8-1 and 8-5.

PROBLEM 8-6
Portray the results of our hypothetical quiz in the form of a table similar to Table 8-1, with the lowest score at the top and the highest score at the bottom. But in addition to the absolute frequency values, include a column showing cumulative absolute frequencies in ascending order from top to bottom.

SOLUTION 8-6

See Table 8-4. Note that the values in the third column, which shows the cumulative absolute frequency, always increase as we go down the table. That is, each number is greater than the one above it. In addition, the largest value is equal to the total number of elements in the statistical group, in this case 130, the number of students in the class.

Table 8-4   Table for Problem 8-6. Note that the cumulative absolute frequency values constantly increase as you read down the table.

Test score	Absolute frequency	Cumulative absolute frequency
0	0	0
1	4	4
2	7	11
3	10	21
4	15	36
5	24	60
6	22	82
7	24	106
8	15	121
9	7	128
10	2	130

PROBLEM 8-7

Render the data from Problem 8-1 in the form of a dual point-to-point graph, showing the lowest score at the left and the highest score at the right on the horizontal scale. Show the absolute frequency values referenced to a vertical scale at the left-hand side of the graph. Show the cumulative absolute fre-quency values as a dashed line, referenced to a vertical scale at the right-hand side of the graph.

SOLUTION 8-7

See Fig. 8-4. To further help in differentiating between the graph of the absolute frequency and the graph of the cumulative absolute frequency, open circles are used to indicate the points corresponding to the cumulative absolute frequency values. The solid black dots and the solid black line are referenced to the left-hand scale; the open circles and the dashed line are referenced to the right-hand scale.

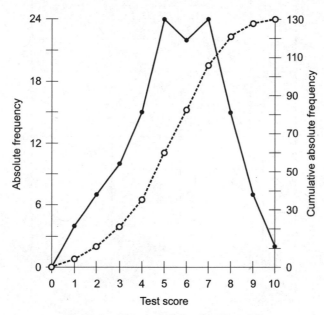

Fig. 8-4.   Illustration for Problem 8-7.

PROBLEM 8-8

What is the mean score for the quiz results in the class we have been discuss-ing in the past few paragraphs?

SOLUTION 8-8

Multiply each score by its absolute frequency, obtaining a set of products. Add up the products and divide the result by the number of papers in the class, in this case 130. Table 8-5 shows the products, along with the cum-ulative sums. Making a table and then double-checking the results can be

Table 8-5 Table for Problem 8-8. The lowest score is at the top and the highest score is at the bottom. The number at the lower right is divided by the number of elements in the population (in this case 130) to obtain the mean, which turns out to be approximately 5.623.

Test score	Absolute frequency	Abs. freq × score	Cum. sum of products
0	0	0	0
1	4	4	4
2	7	14	18
3	10	30	48
4	15	60	108
5	24	120	228
6	22	132	360
7	24	168	528
8	15	120	648
9	7	63	711
10	2	20	731

helpful in situations like this, because errors are easy to make. (If a mistake occurs, it propagates through the rest of the calculation and gets multiplied, worsening the inaccuracy of the final result.) The population mean is 731/130, or approximately 5.623. It can be symbolized μ.

PROBLEM 8-9

What is the median score for the quiz results in the class we have been discussing in the past few paragraphs?

SOLUTION 8-9

Recall the definition of the median from Chapter 2. If the number of elements in a distribution is even, then the median is the value such that half the elements are greater than or equal to it, and half the elements are less than or equal to it. If the number of elements is odd, then the median is the value such that the number of elements greater than or equal to it is the same as the number of elements less than or equal to it. In this case, the "elements" are the test results for each individual in the class. Table 8-6 shows how the median is determined. When the scores of all 130 individual papers are tallied up so they are in order, the scores of the 65th and 66th papers – the two in the

Table 8-6   Table for Problem 8-9. The median can be determined by tabulating the cumulative absolute frequencies.

Test score	Absolute frequency	Cumulative absolute frequency
0	0	0
1	4	4
2	7	11
3	10	21
4	15	36
5	24	60
6 (partial)	5	65
6 (partial)	17	82
7	24	106
8	15	121
9	7	128
10	2	130

middle – are found to be 6 correct. Thus, the median score is 6, because half the students scored 6 or above, and the other half scored 6 or below.

PROBLEM 8-10
What is the mode score for the quiz results in the class we have been discussing in the past few paragraphs?

SOLUTION 8-10
The mode is the score that occurs most often. In this situation there are two such scores, so this distribution is bimodal. The mode scores are 5 and 7.

PROBLEM 8-11
Show the mean, median, and modes as vertical dashed lines in a point-to-point graph similar to the plot of Fig. 8-3.

SOLUTION 8-11
See Fig. 8-5. The mean, median, and modes are labeled, and are all referenced to the horizontal scale.

Variance and Standard Deviation

Let's further analyze the results of the hypothetical 10-question quiz given to the class of 130 students. Again, in verbal form, here are the results:

- 0 correct answers: 0 students
- 1 correct answer: 4 students
- 2 correct answers: 7 students
- 3 correct answers: 10 students
- 4 correct answers: 15 students
- 5 correct answers: 24 students
- 6 correct answers: 22 students
- 7 correct answers: 24 students
- 8 correct answers: 15 students
- 9 correct answers: 7 students
- 10 correct answers: 2 students

PROBLEM 8-12
What is the variance of the distribution of scores for the hypothetical quiz whose results are described above? Round off the answer to three decimal places.

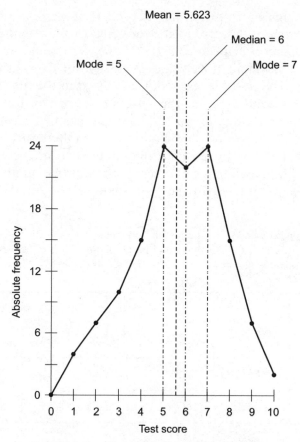

Fig. 8-5. Illustration for Problem 8-11.

SOLUTION 8-12

Let the independent variable be represented as x. There are $n = 130$ individual quiz scores, one for each student. Let's call these individual scores x_i, where i is an *index number* that can range from 1 to 130 inclusive, so there are 130 x_i's ranging from x_1 to x_{130}. Let's call the absolute frequency for each particular numerical score f_j, where j is an index number that can range from 0 to 10 inclusive, so there are 11 f_j's ranging from f_0 to f_{10}. The population mean, μ_p, is approximately 5.623, as we have already found. We are looking for the variance of x, symbolized Var(x). We learned in Chapter 2 that the variance of a set of n values x_1 through x_n with population mean μ is given by the following formula:

$$\text{Var}(x) = (1/n)[(x_1 - \mu)^2 + (x_2 - \mu)^2 + \ldots + (x_n - \mu)^2]$$

It's messy to figure this out in our hypothetical scenario where there are 130 individual test scores, grouped according to the number of students attaining each score. It helps to compile a table as we grind our way through the above formula, and fill in the values as calculations are made. Table 8-7 is our crutch here. The possible test scores are listed in the first (far-left-hand) column. In the second column, the absolute frequency for each score is shown. In the third column, the population mean, 5.623, is subtracted from the score. This gives us a tabulation of the differences between μ and each particular score x_i. In the fourth column, each of these differences is squared. In the fifth column, the numbers in the fourth column are multiplied by the absolute frequencies. In the sixth column, the numbers in the fifth

Table 8-7 Table for Problem 8-12. The population mean, μ_p, is 5.623. The individual test scores are denoted x_i. There are 130 students; i ranges from 1 to 130 inclusive. The absolute frequency for each score is denoted f_j. There are 11 possible scores; j ranges from 0 through 10 inclusive.

Test score x_i	Abs. freq. f_j	$x_i - 5.623$	$(x_i - 5.623)^2$	$f_j (x_i - 5.623)^2$	Cum. sum of $f_j (x_i - 5.623)^2$ values
0	0	−5.623	31.62	0.00	0.00
1	4	−4.623	21.37	85.48	85.48
2	7	−3.623	13.13	91.91	177.39
3	10	−2.623	6.88	68.80	246.19
4	15	−1.623	2.63	39.45	285.64
5	24	−0.623	0.39	9.36	295.00
6	22	0.377	0.14	3.08	298.08
7	24	1.377	1.90	45.60	343.68
8	15	2.377	5.65	84.75	428.43
9	7	3.377	11.40	79.80	508.23
10	2	4.377	19.16	38.32	546.55

column are cumulatively added up. This is the equivalent of summing all the formula terms inside the square brackets, and gives us a grand total of 546.55 at the lower-right corner of the table. This number must be multiplied by $1/n$ (or divided by n) to get the variance. In our case, $n = 130$. Therefore:

$$\text{Var}(x) = 546.55/130$$
$$= 4.204$$

PROBLEM 8-13
What is the standard deviation of the distribution of scores for the quiz? Round off the answer to three decimal places.

SOLUTION 8-13
The standard deviation, symbolized σ, is equal to the square root of the variance. We already know the variance, which is 4.204. The square root of this is found easily using a calculator. For best results, take the square root of the original quotient:

$$\sigma = (546.55/130)^{1/2}$$
$$= 2.050$$

PROBLEM 8-14
Draw a point-to-point graph of the distribution of quiz scores in the situation we've been dealing with in the past few paragraphs, showing the mean, and indicating the range of scores within 1 standard deviation of the mean $(\mu \pm \sigma)$.

SOLUTION 8-14
First, calculate the values of the mean minus the standard deviation $(\mu - \sigma)$ and the mean plus the standard deviation $(\mu + \sigma)$:

$$\mu - \sigma = 5.623 - 2.050 = 3.573$$
$$\mu + \sigma = 5.623 + 2.050 = 7.673$$

These are shown, along with the value of the mean itself, in Fig. 8-6. The scores that fall within this range are indicated by enlarged dots with light interiors. The scores that fall outside this range are indicated by smaller, solid black dots.

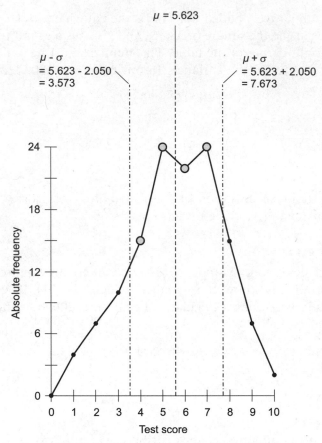

Fig. 8-6.   Illustration for Problem 8-14.

PROBLEM 8-15

In the quiz scenario we've been analyzing, how many students got scores within one standard deviation of the mean? What percentage of the total students does this represent?

SOLUTION 8-15

The scores that fall within the range $\mu \pm \sigma$ are 4, 5, 6, and 7. Referring to the original data portrayed in Table 8-1, we can see that 15 students got 4 answers right, 24 students got 5 answers right, 22 students got 6 answers right, and 24 students got 7 answers right. Therefore, the total number, call it n_σ, of students who got answers within the range $\mu \pm \sigma$ is simply the sum of these:

$$n_\sigma = 15 + 24 + 22 + 24$$
$$= 85$$

The total number of students in the class, n, is equal to 130. To calculate the percentage of students, call it $n_{\sigma\%}$, who scored in the range $\mu \pm \sigma$, we must divide n_σ by n and then multiply by 100:

$$n_{\sigma\%} = 100(n_\sigma/n)$$
$$= 100(85/130)$$
$$= 65.38\%$$

Probability

The next few problems deal with situations and games involving chance. Probability lends itself to "thought experiments" that don't require any material investment except a few sheets of paper, a pen, and a calculator.

PLATONIC DICE

A *Platonic solid* is a geometric solid with flat faces, all of which have identical, regular shape and size. A *regular tetrahedron* has 4 equilateral-triangle faces. A *regular hexahedron* (more often called a cube) has 6 identical square faces. A *regular octahedron* has 8 identical faces, all of which are equilateral triangles.

The following problems involve hypothetical dice that are regular octahedrons, with sides numbered from 1 through 8. Let's call them *octahedral dice*.

PROBLEM 8-16
Suppose you have an octahedral die that is not biased. That is, for any single toss, the probability of the die coming up on any particular face is the same as the probability of its coming up on any other face. (For our purposes, the term "come up" means that a face is oriented facing straight up, so the plane containing it is parallel to the table on which the octahedral die is tossed.) What is the probability that it will come up showing 5 twice in a row? How about 3 times in a row, or 4 times in a row, or n times in a row, where n is a whole number?

SOLUTION 8-16

The probability P_1 that the octahedral die will come up on any particular face (including 5), on any particular toss, is equal to 1/8. That's 0.125 or 12.5%. The probability P_2 that it will come up showing 5 twice in a row is equal to $(1/8)^2$, or 1/64. That's 0.015625 or 1.5625%. The probability P_3 that it will come up showing 5 exactly 3 times in a row is $(1/8)^3$, or 1/512. The probability P_4 that it will come up showing 5 exactly 4 times in a row is $(1/8)^4$, or 1/4096. In general, the probability P_n that the octahedral die will come up showing 5 exactly n times in a row is equal to $(1/8)^n$. Each time we toss the octahedral die, the chance of adding to our consecutive string of 5's is exactly 1 in 8.

PROBLEM 8-17

Imagine a set of octahedral dice, none of which are biased. Suppose you throw a pair of them simultaneously. What's the probability that both octahedral dice will come up showing 3? Now suppose you throw 3 unbiased octahedral dice at the same time. What's the probability that all 3 of them will come up showing 3? How about 4 simultaneously tossed octahedral dice all coming up 3? How about n simultaneously tossed octahedral dice all coming up 3?

SOLUTION 8-17

The progression of probabilities here is the same as that for a single octahedral die tossed over and over. The probability P_1 that the octahedral die will come up on any particular face (including 5), on a single toss, is equal to 1/8. The probability P_2 that a pair of octahedral dice will both come up on any particular face, such as 3, is $(1/8)^2$, or 1/64. The probability P_3 that 3 octahedral dice will come up showing 3 is $(1/8)^3$. The probability P_4 that 4 of them will come up showing 3 is $(1/8)^4$. In general, the probability P_n that n octahedral dice, tossed simultaneously, will all come up showing 3 is equal to $(1/8)^n$.

PROBLEM 8-18

Suppose an election is held for a seat in the state senate. In our hypothetical district, most of the residents are politically liberal. There are two candidates in this election, one of whom is liberal and has been in office for many years, and the other of whom is conservative and has little support. The liberal candidate wins by a landslide, getting 90% of the popular vote. Suppose you pick 5 ballots at random out of various ballot boxes scattered randomly around the district. (This region uses old-fashioned voting methods, despite its liberal slant.) What is the probability that you'll select 5 ballots, all of which indicate votes for the winning candidate?

SOLUTION 8-18

We must be sure we define probability correctly here to avoid the "probability fallacy." We are seeking the proportion of the time that 5 randomly selected ballots will all show votes for the winner, if we select 5 ballots at random on numerous different occasions. For example, if we choose 5 ballots on, say, 1000 different occasions, we want to know how many of these little sneak peeks, as a percentage, will produce 5 ballots all cast for the winner.

Now that we're clear about what we seek, let's convert the winning candidate's percentage to a decimal value. The number 90%, as a proportion, is 0.9. We raise this to the fifth power to get the probability P_5 that 5 randomly selected ballots will all show votes for the winner:

$$P_5 = (0.9)^5$$
$$= 0.9 \times 0.9 \times 0.9 \times 0.9 \times 0.9$$
$$= 0.59049$$
$$= 59.049\%$$

We can round this off to 59%.

PROBLEM 8-19

Suppose you pick n ballots at random out of various ballot boxes scattered randomly around the district in the above scenario, where n is some arbitrary whole number. What is the probability that all the ballots will indicate votes for the winning candidate?

SOLUTION 8-19

Again, we should be clear about what we're seeking. We want to determine the proportion of the time that n randomly selected ballots will all show votes for the winner, if we select n ballots at random a large number of times.

The probability P_n that n randomly selected ballots will all show votes for the winner is equal to 0.9^n, that is, 0.9 raised to the nth power. Expressed as a percentage, the probability $P_{\%n}$ is equal to $(100 \times 0.9^n)\%$. The process of calculating this figure can become tedious for large values of n unless your calculator has an "x to the y power" key. Fortunately, most personal computer calculator programs, when set for scientific mode, have such a key.

PROBLEM 8-20

Generate a table, and plot a graph of $P_{\%n}$ as a function of n, for the scenario described in Problem 8-19. Include values of n from 1 to 10 inclusive.

SOLUTION 8-20

See Table 8-8 and Fig. 8-7. In the graph of Fig. 8-7, the dashed curve indicates the general trend, but the values of the function are confined to the individual points.

Table 8-8 Table for Problems 8-19 and 8-20. The probability $P_{\%n}$ that n randomly selected ballots will all show votes for the winner is equal to $(100 \times 0.9^n)\%$. Probabilities are rounded to the nearest tenth of a percent.

Number of ballots	Probability that all ballots show winning vote
1	90.0%
2	81.0%
3	72.9%
4	65.6%
5	59.0%
6	53.1%
7	47.8%
8	43.0%
9	38.7%
10	34.9%

PROBLEM 8-21

Consider a scenario similar to the one above, except instead of 90% of the popular vote, the winner receives 80% of the popular vote. Suppose you pick n ballots at random out of various ballot boxes scattered randomly around the district in the above scenario, where n is some arbitrary whole number. What is the probability that you'll select n ballots, all of which show votes for the winning candidate?

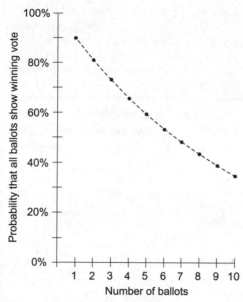

Fig. 8-7. Illustration for Problems 8-19 and 8-20.

SOLUTION 8-21

The probability P_n that n randomly selected ballots will all show votes for the winner is equal to 0.8^n, that is, 0.8 raised to the nth power. Expressed as a percentage, the probability $P_{\%n}$ is equal to $(100 \times 0.8^n)\%$.

PROBLEM 8-22

Generate a table, and plot a graph of $P_{\%n}$ as a function of n, for the scenario described in Problem 8-21. Include values of n from 1 to 10 inclusive.

SOLUTION 8-22

See Table 8-9 and Fig. 8-8. In the graph of Fig. 8-8, the dashed curve indicates the general trend, but the values of the function are confined to the individual points.

PROBLEM 8-23

Plot a dual graph showing the scenarios of Problems 8-19 and 8-21 together, thereby comparing the two.

SOLUTION 8-23

See Fig. 8-9. The dashed curves indicate the general trends, but the values of the functions are confined to the individual points.

Table 8-9   Table for Problems 8-21 and 8-22. The probability $P_{\%n}$ that n randomly selected ballots will all show votes for the winner is equal to $(100 \times 0.8^n)\%$. Probabilities are rounded to the nearest tenth of a percent.

Number of ballots	Probability that all ballots show winning vote
1	80.0%
2	64.0%
3	51.2%
4	41.0%
5	32.8%
6	26.2%
7	21.0%
8	16.8%
9	13.4%
10	10.7%

Data Intervals

The following problems involve data intervals including quartiles, deciles, percentiles, and straight fractional portions. Let's consider an example involving climate change. (The following scenario is fictitious, and is for illustrative purposes only. It shouldn't be taken as actual history.)

THE DISTRIBUTION

Suppose you want to know if the average temperature in the world has increased over the last 100 years. You obtain climate data for many cities

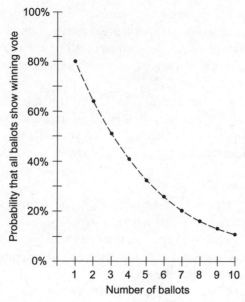

Fig. 8-8. Illustration for Problems 8-21 and 8-22.

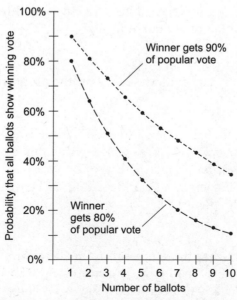

Fig. 8-9. Illustration for Problem 8-23.

and towns scattered throughout the world. You are interested in one figure for each location: the average temperature over the course of last year, versus the average temperature over the course of the year one century ago. The term "a century ago" or "100 years earlier" means "100 years before this year (99 years before last year)."

In order to calculate a meaningful figure for any given locale, you compare last year's average temperature t, expressed in degrees Celsius (°C), with the average annual temperature s during the course of the year a century ago. You figure the temperature change, T, in degrees Celsius simply as follows:

$$T = t - s$$

If $t < s$, then T is negative, indicating that the temperature last year was lower than the temperature a century ago. If $t = s$, then $T = 0$, meaning that the temperature last year was the same as the temperature a century ago. If $t > s$, then T is positive, meaning that the temperature last year was higher than the temperature a century ago.

Now imagine that you have obtained data for so many different places that generating a table is impractical. Instead, you plot a graph of the number of locales that have experienced various average temperature changes between last year and a century ago, rounded off to the nearest tenth of a degree Celsius. Suppose the resulting smoothed-out curve looks like the graph of Fig. 8-10. We could generate a point-by-point graph made up of many short, straight line segments connecting points separated by 0.1°C on the horizontal scale, but that's not what we've done here. Instead, Fig. 8-10 is a smooth, continuous graph obtained by curve fitting.

PROBLEM 8-24

What do the points $(-2,18)$ and $(+2.8,7)$ on the graph represent?

SOLUTION 8-24

These points tell us that there are 18 locales whose average annual temperatures were lower last year by 2°C as compared with a century ago, as shown by the point $(-2,18)$, and that there are 7 locales whose average annual temperatures were higher last year by 2.8°C as compared with a century ago, as shown by the point $(+2.8,7)$.

PROBLEM 8-25

Suppose we are told that the temperature in the town of Thermington was higher last year by 1.3°C than it was a century ago. This fact is indicated by the vertical, dashed line in the graph of Fig. 8-10. Suppose L represents the proportion of the area under the curve (but above the horizontal axis show-

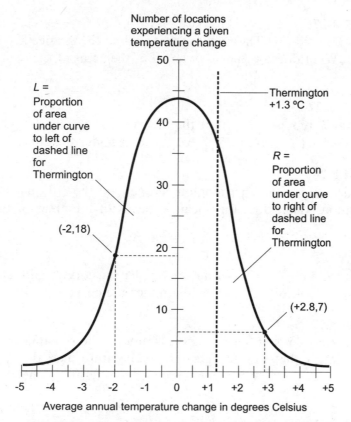

Fig. 8-10.   Illustration for Problems 8-24 through 8-30.

ing temperatures) to the left of this dashed line, and *R* represents the proportion of the area under the curve to the right of the dashed line. What can be
said about *L* and *R*?

SOLUTION 8-25
The sum of *L* and *R* is equal to 1. If *L* and *R* are given as percentages, then
$L + R = 100\%$.

PROBLEM 8-26
Suppose we are told that Thermington is exactly at the 81st percentile point
in the distribution. What does this mean in terms of the areas of the regions
L and *R*?

SOLUTION 8-26
It means that *L* represents 81% of the area under the curve, and therefore
that *R* represents $100\% - 81\%$, or 19%, of the area under the curve.

PROBLEM 8-27
Suppose we are told that Thermington is exactly at the 8th decile point in the distribution. What does this mean in terms of the areas of the regions L and R?

SOLUTION 8-27
It means that L represents 8/10 of the area under the curve, and therefore that R represents $1 - 8/10$, or 2/10, of the area under the curve.

PROBLEM 8-28
Suppose we are told that Thermington is exactly at the 3rd quartile point in the distribution. What does this mean in terms of the areas of the regions L and R?

SOLUTION 8-28
It means that L represents 3/4 of the area under the curve, and therefore that R represents $1 - 3/4$, or 1/4, of the area under the curve.

PROBLEM 8-29
Suppose we are told that Thermington is among the one-quarter of towns in the experiment that saw the greatest temperature increase between last year and 100 years ago. What does this mean in terms of the areas L and R?

SOLUTION 8-29
The statement of this problem is ambiguous. It could mean either of two things:

1. We are considering all the towns in the experiment.
2. We are considering only those towns in the experiment that witnessed increases in temperature.

If we mean (1) above, then the above specification means that L represents more than 3/4 of the area under the curve, and therefore that R represents less than 1/4 of the area under the curve. If we mean (2) above, we can say that R represents less than 1/4, or 25%, of the area under the curve to the right of the vertical axis (the axis in the center of the graph showing the number of locations experiencing a given temperature change); but we can't say anything about L unless we know more about the distribution.

It looks like the curve in Fig. 8-10 is symmetrical around the vertical axis. In fact, it's tempting to think that the curve is a normal distribution. But we haven't been told that this is the case. We mustn't assume it without proof. Suppose we run the data through a computer and determine that the curve is a normal distribution. Then if (2) above is true, L represents more than 7/8 of the total area under the curve, and R represents less than 1/8 of the total area under the curve.

PROBLEM 8-30

Suppose that the curve in Fig. 8-10 represents a normal distribution, and that Thermington happens to lie exactly one standard deviation to the right of the vertical axis. What can be said about the areas L and R in this case?

SOLUTION 8-30

Imagine the "sister city" of Thermington, a town called Frigidopolis. Suppose Frigidopolis was cooler last year, in comparison to 100 years ago, by exactly the same amount that Thermington was warmer. This situation is shown graphically in Fig. 8-11. Because Thermington corresponds to a point (or dashed vertical line) exactly one standard deviation to the right of the

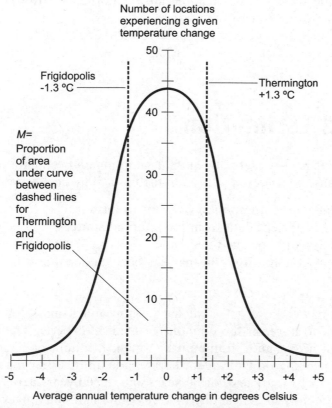

Fig. 8-11.   Illustration involving the solution to Problem 8-30.

vertical axis, Frigidopolis can be represented by a point (or dashed vertical line) exactly one standard deviation to the left of the vertical axis.

The mean (μ) in Fig. 8-11 happens to coincide with the vertical axis, or the point where the temperature change is 0. Recall the empirical rule concerning standard deviation (σ) and normal distributions. We learned about this in Chapter 3. The empirical rule applies directly to this problem. The vertical dashed line on the left, representing Frigidopolis, is $-\sigma$ from μ. The vertical dashed line on the right, representing Thermington, is $+\sigma$ from μ. The empirical rule tells us that the proportion of the area under the curve between these two dashed lines is 68% of the total area under the curve, because these two dashed vertical lines represent that portion of the area within $\pm\sigma$ of μ. The proportion of the area under the curve between the vertical axis and either dashed line is half this, or 34%.

The fact that Fig. 8-11 represents a normal distribution also tells us that exactly 50% of the total area under the curve lies to the left of μ, and 50% of the area lies to the right of μ, if we consider the areas extending indefinitely to the left or the right. This is true because a normal distribution is always symmetrical with respect to the mean.

Knowing all of the above, we can determine that $L = 50\% + 34\% = 84\%$. That means $R = 100\% - 84\% = 16\%$.

Sampling and Estimation

Here are some problems involving data sampling and estimation. Recall the steps that must be followed when conducting a statistical experiment:

1. Formulate the question(s) we want to answer.
2. Gather sufficient data from the required sources.
3. Organize and analyze the data.
4. Interpret the information that has been gathered and organized.

PROBLEM 8-31
Most of us have seen charts that tell us how much mass (or weight) our bodies ought to have, in the opinions of medical experts. The mass values are based on height and "frame type" (small, medium, or large), and may also vary depending on your age. Charts made up this year may differ from those made up 10 years ago, or 30 years ago, or 60 years ago.

Let's consider hypothetical distributions of human body mass as functions of body height, based on observations of real people rather than on idealized theories. Imagine that such distributions are compiled for the whole world, and are based solely on how tall people are; age, ethnicity, gender, and all other factors are ignored. Imagine that we end up with a set of many graphs, one plot for each height in centimeters, ranging from the shortest person in the world to the tallest, showing how massive people of various heights actually are.

Suppose we obtain a graph (Fig. 8-12, hypothetically) showing the mass distribution of people in the world who are 170 centimeters tall, rounded to the nearest centimeter. Suppose we round each individual person's mass to the nearest kilogram. Further suppose that this is a normal distribution. Therefore, the mean, median, and mode are all equal to 55 kilograms. We

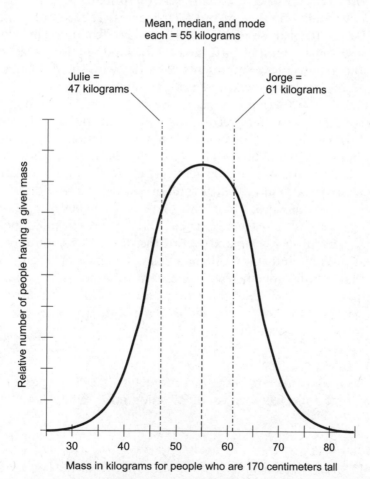

Fig. 8-12.   Illustration for Problems 8-31 and 8-32.

obviously can't put every 170-centimeter-tall person in the world on a scale, measure his or her mass, and record it! Suggest a sampling frame that will ensure that Fig. 8-12 is a good representation of the actual distribution of masses for people throughout the world who are 170 centimeters tall. Suggest another sampling frame that might at first seem good, but that in fact is not.

SOLUTION 8-31

Recall the definition of a sampling frame: a set of elements from within a population, from which a sample is chosen. A sampling frame is a representative cross-section of a population.

Let's deal with the not-so-good idea first. Suppose we go to every recognized country in the world, randomly select 100 males and 100 females, all 170 centimeters tall, and then record their masses. Then, as our samples, we randomly select 10 of these males and 10 of these females. This won't work well, because some countries have much larger populations than others. It will skew the data so it over-represents countries with small populations and under-represents countries with large populations.

If we modify the preceding scheme, we can get a fairly good sampling frame. We might go to every recognized country in the world, and select a number of people in that country based on its population in thousands. So if a country has, say, 10 million people, we would select, as nearly at random as possible, 10,000 males and 10,000 females, all 170 centimeters tall, as our sampling frame. If a country has 100 million people, we would select 100,000 males and 100,000 females. If a country has only 270,000 people, we would select 270 males and 270 females, each 170 centimeters tall. Then we put every tenth person on a scale and record the mass. If a country has so few people that we can't find at least 10 males and 10 females 170 centimeters tall, we lump that country together with one or more other small countries, and consider them as a single country. From this, we obtain the distribution by using a computer to plot all the individual points and run a curve-fitting program to generate a smooth graph.

PROBLEM 8-32

Suppose Jorge and Julie are both 170 centimeters tall. Jorge has a mass of 61 kilograms, and Julie has a mass of 47 kilograms. Where are they in the distribution?

SOLUTION 8-32

Figure 8-12 shows the locations of Jorge and Julie in the distribution, relative to other characteristics and the curve as a whole.

PROBLEM 8-33

Suppose that we define the "typical human mass" for a person 170 centimeters tall to be anything within 3 kilograms either side of 55 kilograms. Also suppose that we define ranges called "low typical" and "high typical," representing masses that are more than 3 kilograms but less than 5 kilograms either side of 55 kilograms. Illustrate these ranges in the distribution.

SOLUTION 8-33

Figure 8-13 shows these ranges relative to the whole distribution.

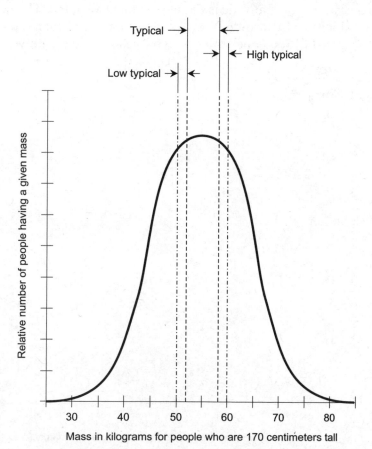

Fig. 8-13. Illustration for Problem 8-33.

Hypotheses, Prediction, and Regression

The following several problems involve a hypothetical experiment in which the incidence of a mystery illness, Syndrome X, is scrutinized. Twenty groups of 100 people are selected according to various criteria. Then the percentage of people in each group who exhibit Syndrome X is recorded. Graphs are compiled, and the data is analyzed. This example, like all the others in this chapter, is based on real-world possibilities, but otherwise it's entirely make-believe.

PROBLEM 8-34

Suppose we test 100 randomly selected people living at locations in 20 different latitudes (for a total of 2000 people in the experiment). We render the results as a scatter plot. Latitude is the independent variable. The percentage of people exhibiting Syndrome X is the dependent variable. The result is shown in Fig. 8-14. Someone states a hypothesis: "If you move closer to,

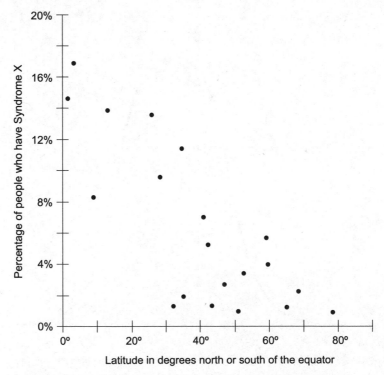

Fig. 8-14.   Illustration for Problems 8-34, 8-35, 8-36, and 8-39.

or farther from, the equator, your risk of developing Syndrome X does not change." What sort of hypothesis is this? Someone else says, "Look at Fig. 8-14. Obviously, people who live close to the equator get Syndrome X more often than people who live far from the equator. I believe that if you move closer to the equator, your risk of developing Syndrome X will increase." What sort of hypothesis is this? A third person says, "The scatter plot shows that a greater proportion of people who live close to the equator have Syndrome X, as compared with people who live far from the equator. But this does not logically imply that if you move closer to the equator, then you run a greater risk of developing Syndrome X than if you stay here or move farther from the equator." What sort of hypothesis is this?

SOLUTION 8-34
The first hypothesis is an example of a null hypothesis. The second and third hypotheses are examples of alternative hypotheses.

PROBLEM 8-35
Provide an argument that can be used to support the first hypothesis in Problem 8-34.

SOLUTION 8-35
The first hypothesis is this: "If you move closer to, or farther from, the equator, your risk of developing Syndrome X will not change." The scatter plot of Fig. 8-14 shows that people who live close to the equator have Syndrome X in greater proportion than people who live far from the equator. But it is an oversimplification to say that the latitude of residence, all by itself, is responsible for Syndrome X. The syndrome might be preventable by taking precautions that most people who live near the equator don't know about, or in which they don't believe, or that their governments forbid them to take. If you live in Amsterdam and are knowledgeable about Syndrome X, you might adjust your lifestyle or take a vaccine so that, if you move to Singapore, you will bear no greater risk of contracting Syndrome X than you have now.

PROBLEM 8-36
How can the first hypothesis in Problem 8-34 be tested?

SOLUTION 8-36
In order to discover whether or not moving from one latitude to another affects the probability that a person will develop Syndrome X, it will be necessary to test a large number of people who have moved from various specific places to various other specific places. This test will be more complex

and time-consuming than the original experiment. Additional factors will enter in, too. For example, we will have to find out how long each person has lived in the new location after moving, and how much traveling each person does (for example, in conjunction with employment). The extent, and not only the direction, of the latitude change will also have to be taken into account. Is there a difference between moving from Amsterdam to Singapore, as compared with moving from Amsterdam to Rome? Another factor is the original residence latitude. Is there a difference between moving from Rome to Singapore, as compared with moving from Amsterdam to Singapore?

PROBLEM 8-37
Figure 8-15 shows a scatter plot of data for the same 20 groups of 100 people that have been researched in our hypothetical survey involving Syndrome X. But instead of the latitude in degrees north or south of the equator, the altitude, in meters above sea level, is the independent variable. What does this graph tell us?

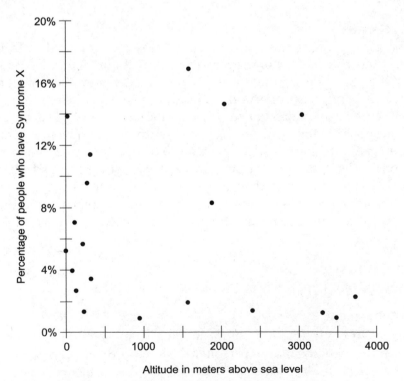

Fig. 8-15.   Illustration for Problems 8-37, 8-38, and 8-40.

SOLUTION 8-37

It is difficult to see any correlation here. Some people might see a weak negative correlation between the altitude of a place above sea level and the proportion of the people exhibiting Syndrome X. But other people might see a weak positive correlation because of the points in the upper-right portion of the plot. A computer must be used to determine the actual correlation, and when it is found, it might turn out to be so weak as to be insignificant.

PROBLEM 8-38

Suppose someone comes forward with a hypothesis: "If you move to a higher or lower altitude above sea level, your risk of developing Syndrome X does not change." What sort of hypothesis is this? Someone else says, "It seems to me that Fig. 8-15 shows a weak, but not a significant, correlation between altitude and the existence of Syndrome X in the resident population. But I disagree with you concerning the hazards involved with moving. There might be factors that don't show up in this data, even if the correlation is equal to 0; and one or more of these factors might drastically affect your susceptibility to developing Syndrome X if you move much higher up or lower down, relative to sea level." What sort of hypothesis is this?

SOLUTION 8-38

The first hypothesis is a null hypothesis. The second hypothesis is an alternative hypothesis.

PROBLEM 8-39

Estimate the position of the line of least squares for the scatter plot showing the incidence of Syndrome X versus the latitude north or south of the equator (Fig. 8-14).

SOLUTION 8-39

Figure 8-16 shows a "good guess" at the line of least squares for the points in Fig. 8-14.

PROBLEM 8-40

Figure 8-17 shows a "guess" at a regression curve for the points in Fig. 8-15, based on the notion that the correlation is weak, but negative. Is this a "good guess"? If so, why? If not, why not?

SOLUTION 8-40

Figure 8-17 is not a "good guess" at a regression curve for the points in Fig. 8-15. There is no such thing as a "good guess" here. The correlation is weak at best, and its nature is uncertain in the absence of computer analysis.

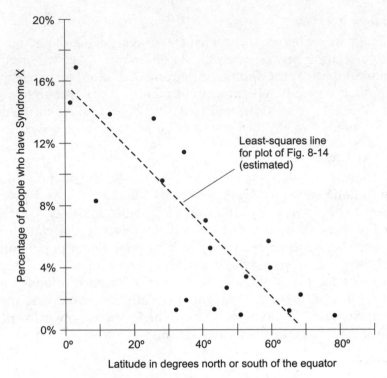

Fig. 8-16. Illustration for Problem 8-39.

Correlation and Causation

As we learned in Chapter 7, correlation between two variables does not imply a direct cause-and-effect relation. Here are some problems involving correlation and causation.

TWO MODERN PLAGUES

During the last few decades of the 20th century and into the first years of the 21st century, there was a dramatic increase in the incidence of adult-onset diabetes (let's call it AOD for short) in the United States. This is a syndrome in which the body develops problems regulating the amount of glucose in the blood. (Glucose is a simple form of sugar and an important body fuel.) During the same period of time, there was also an increase in the incidence of obesity (overweight). Scientists and physicians have long suspected that

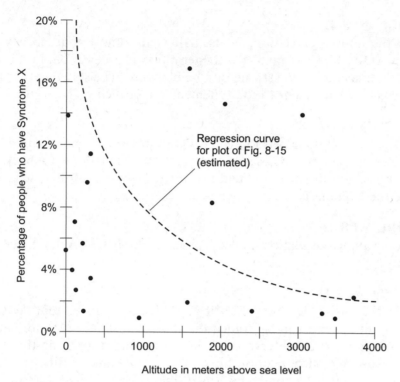

Fig. 8-17. Illustration for Problem 8-40.

there are cause-and-effect relationships at work here, but the exact mechanisms have been under debate.

PROBLEM 8-41

It has been found that overweight adults are more likely to have AOD than adults who are not overweight. In fact, if a person is randomly selected from the United States population, the likelihood of that person having AOD increases as the extent of the obesity increases. What sort of correlation is this?

SOLUTION 8-41

Obesity is positively correlated with AOD. To be more precise, let M_n be the normal mass in kilograms for a person of a certain height, gender, and age, and let M_a be the person's actual mass in kilograms. Then the probability that a randomly selected person has AOD is positively correlated with the value of M_a/M_n.

PROBLEM 8-42

Does the above correlation, in itself, logically imply that obesity directly causes AOD? Keep in mind that the term *logical implication*, in this context, means implication in the strongest possible sense. The statement "*P* logically implies *Q*" is equivalent to the statement "If *P*, then *Q*."

SOLUTION 8-42

No. The "conventional wisdom" holds that obesity is a contributing factor to the development of AOD, and research has been done to back up this hypothesis. But the mere existence of the correlation, all by itself, does not prove the hypothesis.

PROBLEM 8-43

Does the above correlation, in itself, logically imply that AOD directly causes obesity?

SOLUTION 8-43

Again, the answer is "No." Few if any scientists believe that AOD causes obesity, although it can be argued that this idea should get more attention. The correlation is an observable phenomenon, but if we claim that there is a direct cause-and-effect relationship between AOD and obesity, we must conduct research that demonstrates a plausible reason, a *modus operandi*.

PROBLEM 8-44

Does the above correlation, in itself, logically imply that both AOD and obesity are caused by some other factor?

SOLUTION 8-44

The answer is "No" yet again, for the same reasons as those given above. In order to conclude that there is a cause-and-effect relationship of any kind between or among variables, good reasons must be found to support such a theory. But no particular theory logically follows from the mere existence of the correlation.

Most scientists seem to agree that there is a causative factor involved with both AOD and obesity: the consumption of too much food! It has been found that people who overeat are more likely to be obese than people who don't, and the recent increase in AOD has also been linked to overeating. But even this is an oversimplification. Some scientists believe that the overconsumption of certain types of food is more likely to give rise to AOD than the overconsumption of other types of food. There are also sociological, psychological, and even political issues involved. It has been suggested that the stress of modern living, or the presence of industrial

pollutants, could give rise to AOD. And who knows that the incidence of AOD might not be influenced by such unsuspected factors as exposure to electromagnetic fields?

PROBLEM 8-45

We have now seen that the correlation between obesity and AOD, all by itself, does not logically imply any particular cause-and-effect relationship. Does this mean that whenever we see a correlation, be it weak or strong or positive or negative, we must conclude that it is nothing more than a coincidence?

SOLUTION 8-45

Once again, the answer is "No." When we observe a correlation, we should resist the temptation to come to any specific conclusion about cause-and-effect in the absence of supporting research or a sound theoretical argument. We should be skeptical, but not closed-minded. We should make every possible attempt to determine the truth, without letting personal bias, peer pressure, economic interests, or political pressures cloud our judgment. There is often a cause-and-effect relationship (perhaps more than one!) behind a correlation, but we must also realize that coincidences can and do occur.

Quiz

Refer to the text in this chapter if necessary. A good score is 8 correct. Answers are in the back of the book.

1. Table 8-10 shows the results of a hypothetical quiz given to a class of 130 students. What is wrong with this table?
    (a) The absolute frequency values don't add up to 130.
    (b) The cumulative absolute frequency values are not added up correctly.
    (c) There exists no mean.
    (d) There exists no median.

2. Imagine a town in the Northern Hemisphere. In this town, the average monthly rainfall is much greater in the winter than in the summer; the spring and fall months are wetter than the summer months but drier than the winter months. The average monthly temperature is much higher in the summer than in the winter; the spring and fall months are cooler than the summer months but warmer than the winter months.

Table 8-10   Table for Quiz Question 1.

Test score	Absolute frequency	Cumulative absolute frequency
0	2	2
1	5	7
2	10	17
3	10	27
4	15	36
5	22	60
6	22	82
7	21	106
8	13	121
9	7	128
10	3	130

The correlation between average monthly temperature and average monthly precipitation is
(a) positive
(b) negative
(c) 0
(d) impossible to express without more information

3. Suppose that the increase in the incidence of adult-onset diabetes (AOD) in recent years was decreasing, even though the incidence of obesity was increasing. This would logically imply that
(a) weight loss causes AOD
(b) weight gain prevents AOD
(c) AOD causes weight loss
(d) none of the above

4. What is the probability that an unbiased octahedral (8-faceted) die will, if tossed 4 times in a row, come up in the sequence 2, 4, 6, 8, in that order?
    (a) 1 in 1024
    (b) 1 in 2048
    (c) 1 in 4096
    (d) 1 in 8192

5. Refer again to the Syndrome X scenario discussed in Problems 8-34 through 8-40. Figure 8-18 is a scatter plot showing the incidence of Syndrome X versus the average annual precipitation in centimeters at the location where each person lives. This plot indicates that there is
    (a) no correlation between the incidence of Syndrome X and the average annual precipitation
    (b) positive correlation between the incidence of Syndrome X and the average annual precipitation

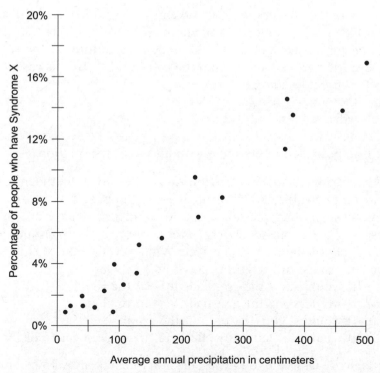

Fig. 8-18. Illustration for Quiz Questions 5 and 6.

(c) negative correlation between the incidence of Syndrome X and the
average annual precipitation
(d) a cause–effect relationship between the incidence of Syndrome X
and the average annual precipitation

6. In Fig. 8-18, the least-squares line is not shown, but if it were shown, it
would
(a) be horizontal
(b) be vertical
(c) ramp downward as you move toward the right
(d) ramp upward as you move toward the right

7. Suppose you toss an unbiased octahedral die 4 times in a row and you
get the sequence 1, 2, 3, 4, in that order. What is the probability that, if
you toss the die a fifth time, it will come up showing 5?
(a) 1 in 8
(b) 1 in 16
(c) 1 in 32
(d) 1 in 64

8. Refer to the 10-question quiz taken by 130 students, and discussed in
Problems 8-1 through 8-15. Examine Fig. 8-6. Imagine that the quiz is
given to another group of 130 students, resulting in the same mean
score (μ) but a smaller standard deviation (σ). In this case, the range
between $\mu - \sigma$ and $\mu + \sigma$ is
(a) the same as it is in Fig. 8-6
(b) wider than it is in Fig. 8-6
(c) narrower than it is in Fig. 8-6
(d) impossible to determine without more information

9. Refer again to the 10-question quiz taken by 130 students, and discussed in Problems 8-1 through 8-15. Suppose a controversy arises
over one of the questions, sparking intense debate and ultimately
causing the professor to give everyone credit for the question, as if
they all got the answers correct. What effect would this have on the
general shape of the graph shown in Fig. 8-6?
(a) It would skew the graph slightly to the right.
(b) It would skew the graph slightly to the left.
(c) It would slightly flatten out the graph.
(d) It would not have any effect on the general shape of the graph.

10. Refer one last time to the 10-question quiz taken by 130 students, and
discussed in Problems 8-1 through 8-15. Suppose that all the scores

were 5 correct, 6 correct, or 7 correct; no one got less than 5 correct nor more than 7 correct. What effect would this have on the general shape of the graph shown in Fig. 8-6?

(a) It would not have any effect on the general shape of the graph.
(b) It would cause the graph to be more sharply peaked.
(c) It would cause the graph to be more flattened out.
(d) There is no way to know, without more information, what would happen to the general shape of the graph.

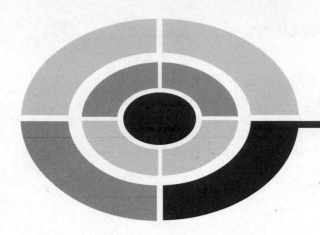

Test: Part Two

Do not refer to the text when taking this test. You may draw diagrams or use a calculator if necessary. A good score is at least 45 correct. Answers are in the back of the book. It's best to have a friend check your score the first time, so you won't memorize the answers if you want to take the test again.

1. The process that is used to draw conclusions on the basis of data and hypotheses is called
   (a) the null hypothesis
   (b) a scenario
   (c) inference
   (d) regression
   (e) correlation

2. Table Test 2-1 shows an example of
   (a) sampling without replacement
   (b) sampling in ascending order
   (c) sampling in descending order
   (d) sampling of primary elements
   (e) sampling of means

Table Test 2-1 Table for Part Two Test Question 2.

Before sampling	Element deleted	After sampling
{ℵ, ♣, ℑ, ♦, ℜ, ♥, ℘, ♠, ∇, Φ}	♦	{ℵ, ♣, ℑ, ℜ, ♥, ℘, ♠, ∇, Φ}
{ℵ, ♣, ℑ, ℜ, ♥, ℘, ♠, ∇, Φ}	℘	{ℵ, ♣, ℑ, ℜ, ♥, ♠, ∇, Φ}
{ℵ, ♣, ℑ, ℜ, ♥, ♠, ∇, Φ}	Φ	{ℵ, ♣, ℑ, ℜ, ♥, ♠, ∇}
{ℵ, ♣, ℑ, ℜ, ♥, ♠, ∇}	♣	{ℵ, ℑ, ℜ, ♥, ♠, ∇}
{ℵ, ℑ, ℜ, ♥, ♠, ∇}	ℑ	{ℵ, ℜ, ♥, ♠, ∇}
{ℵ, ℜ, ♥, ♠, ∇}	♥	{ℵ, ℜ, ♠, ∇}
{ℵ, ℜ, ♠, ∇}	ℵ	{ℜ, ♠, ∇}
{ℜ, ♠, ∇}	♠	{ℜ, ∇}
{ℜ, ∇}	∇	{ℜ}
{ℜ}	ℜ	∅

3. In a normal distribution, the width of the confidence interval is related to
   (a) the difference between the mean and the mode
   (b) the difference between the mean and the median
   (c) the number of modes either side of the standard deviation
   (d) the number of standard deviations either side of the mean
   (e) none of the above

4. Suppose an experiment is conducted in which people's average daily caloric intake is plotted to get a distribution. The intent is to get a good representative sampling of the entire population of the United States. Errors in this experiment might arise because of
   (a) uncertain data on the caloric content of foods
   (b) sloppy reporting or monitoring of people's food intake
   (c) a small number of people sampled
   (d) the gathering of data from cities but not from rural areas
   (e) any of the above

5. Imagine that the above experiment is conducted, and someone says on the basis of the results, "A typical person in the United States consumes between 2700 and 3300 calories a day." This is the equivalent of saying:

(a) "A typical person in the United States has a daily caloric consumption of 3000±10%."

(b) "A typical person in the United States has a daily caloric consumption of 3000±3%."

(c) "A typical person in the United States has a daily caloric consumption of 3000±1%."

(d) "A typical person in the United States has a daily caloric consumption of 3000±0.3%."

(e) "A typical person in the United States has a daily caloric consumption of 3000±0.1%."

6. Imagine a town in a southern U.S. city. There is a computer store in this town. The owner of the store believes there is a correlation between the temperature and the number of computers sold per day. She looks back over the past year's sales records and determines the number of computers sold on 12 different days (one in each month of the year). Then she goes to the Internet and finds the average temperatures (high plus low divided by 2) for each day in her town. When she places the resulting points on a scatter plot, she gets Fig. Test 2-1. This suggests

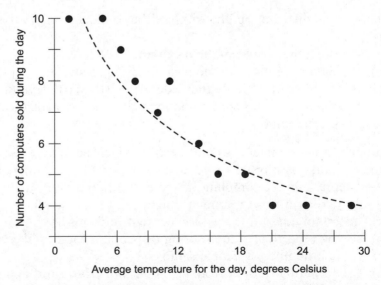

Fig. Test 2-1.   Illustration for Part Two Test Questions 6 and 7.

(a) that there is no correlation between the average temperature on a given day and the number of computers sold by her store on that day

(b) that there is a negative correlation between the average temperature on a given day and the number of computers sold by her store on that day

(c) that there is a positive correlation between the average temperature on a given day and the number of computers sold by her store on that day

(d) that more information is needed to figure out if there is a correlation between the average temperature on a given day and the number of computers sold by her store on that day

(e) no useful information whatsoever

7. The dashed line in Fig. Test 2-1 is an estimate of the
   (a) line of least squares
   (b) regression curve
   (c) correlation ratio
   (d) standard deviation
   (e) variance

8. Imagine that the points in a scatter plot are arranged in a perfect circle, evenly spaced all the way around. What is the orientation of the least-squares line?
   (a) It is horizontal.
   (b) It is vertical.
   (c) It ramps up as you move toward the right.
   (d) It ramps down as you move toward the right.
   (e) It is impossible to determine because infinitely many different lines can, in theory, be defined as the least-squares line.

9. Imagine that the points in a scatter plot are arranged in a perfect circle, evenly spaced all the way around. What is the correlation between the variables in this case?
   (a) 0
   (b) Something between 0 and +1
   (c) +1
   (d) Something between −1 and 0
   (e) −1

10. A hypothesis is always
    (a) an assumption
    (b) provable to be true

(c) provable to be false

(d) a certainty

(e) a logical process

11. Imagine that a distribution is not a normal distribution. However, the sampling distribution of means approaches a normal distribution as the sample size increases. This statement

(a) is patently false. As the sample size increases, the sampling distribution of means becomes less and less like a normal distribution

(b) is true only if the original distribution is almost a normal distribution

(c) is true only for small populations

(d) is true only for large populations

(e) is true because of the Central Limit Theorem

12. Imagine that you want to determine the quantitative effect (if any) that salt consumption has on people's blood pressure. You interview 100 people from each continent in the world, ask them how much salt they eat, and then measure their blood pressures with instruments obtained from, and calibrated by, a respected medical school. Which of the following possible flaws in this experiment should you be the least concerned about?

(a) People might not accurately know how much salt they eat.

(b) The blood pressure measuring equipment might not show correct readings.

(c) The sampling frame is too small.

(d) The sample might be biased relative to such factors as age, gender, or ethnicity.

(e) You need not be concerned about any of the factors (a), (b), (c), or (d).

13. Table Test 2-2 illustrates the average monthly temperatures and rainfall amounts for an imaginary city in the Southern Hemisphere called Rio de Antonio. From this table it is apparent that

(a) there is no correlation between the average monthly temperature and the average monthly rainfall

(b) there is a significant correlation between the average monthly temperature and the average monthly rainfall

(c) there is no correlation between the time of year and the average monthly rainfall

(d) there is a significant correlation between the time of year and the average monthly rainfall

Table Test 2-2 Table for Part Two Test Questions 13 and 14.

Month	Average temperature, degrees Celsius	Average rainfall, centimeters
January	22.1	10.4
February	23.0	10.2
March	20.4	8.3
April	17.6	7.8
May	15.3	6.2
June	12.7	2.3
July	10.8	1.3
August	11.6	1.5
September	14.9	2.7
October	15.6	4.6
November	17.5	7.6
December	19.8	10.5

(e) more than one of the above

14. Refer to Table Test 2-2 and Fig. Test 2-2. Which of the graphs in Fig. Test 2-2 most nearly represents the line of least squares for a scatter plot of the relationship between the average monthly temperature and the average monthly rainfall for Rio de Antonio? Assume that on each plot, temperatures increase as you move to the right along the horizontal scale, and rainfall amounts increase as you move up along the vertical scale.
    (a) A
    (b) B

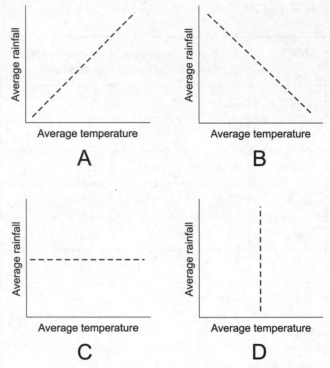

Fig. Test 2-2. Illustration for Part Two Test Question 14.

    (c) C
    (d) D
    (e) None of them

15. Imagine that a correlation of $r_i = -0.500$ is found between the average number of minutes per day people use the Internet and the average number of minutes per day they spend reading books. The time people spend on the Internet is the independent variable, and the time they spend reading books is the dependent variable. If the independent and dependent variables are interchanged, what is the correlation r_r between them?
    (a) $r_r = -0.500$
    (b) $r_r = 0$
    (c) $r_r = +0.500$
    (d) $r_r = -2.00$
    (e) $r_r = +2.00$

16. In a bar graph, what is the advantage of showing the numbers (if there's room) at the ends or the tops of the bars?

(a) It makes the graph appear less cluttered.
(b) It eliminates bias in the experiment.
(c) It tells the observer exactly how tall or wide the bars are.
(d) It turns the graph into a point-to-point plot.
(e) There is no advantage in doing this.

17. The possible range of correlation values between two variables is
    (a) −50% to +50%
    (b) −100% to 0%
    (c) −100% to +100%
    (d) 0% to 100%
    (e) none of the above

18. The level of significance is the probability that the null hypothesis will turn out to be true
    (a) after it has been rejected
    (b) after it has been accepted
    (c) after all the alternative hypotheses have been rejected
    (d) provided all the alternative hypotheses are also true
    (e) provided some of the alternative hypotheses are also true

19. The level of significance can be expressed as
    (a) a value less than −1
    (b) a value between −1 and 1
    (c) a value between −1 and 0
    (d) a value between 0 and 1
    (e) a value greater than 1

20. If data are already available and all a statistician has to do is organize it and analyze it, then it is called
    (a) analyzed source data
    (b) secondary source data
    (c) estimated source data
    (d) correlated source data
    (e) prepared source data

21. Experimental defect error can be caused by
    (a) replacing elements when they should not be replaced
    (b) failing to replace elements when they should be replaced
    (c) attempting to compensate for factors that don't have any real effect
    (d) a sample that is not large enough
    (e) any of the above

22. What is the technical term for the type of graph shown in Fig. Test 2-3?
    (a) A curvature plot.
    (b) A relational plot.
    (c) A scatter plot.
    (d) A correlative plot.
    (e) A least-squares plot.

23. What can be said about the correlation r between Parameter X and Parameter Y in Fig. Test 2-3?
    (a) $-1 < r < 0$
    (b) $0 < r < +1$
    (c) $r = -1$
    (d) $r = +1$
    (e) $r = 0$

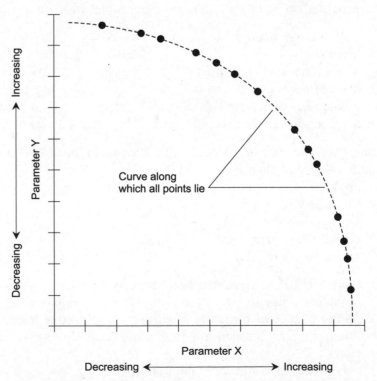

Fig. Test 2-3. Illustration for Part Two Test Questions 22 through 24.

24. Suppose a computer were used in an attempt to locate the least-squares line in Fig. Test 2-3. What would be the orientation of the resulting line?
    (a) It would ramp downward as you move toward the right.
    (b) It would ramp upward as you move toward the right.
    (c) It would be vertical.
    (d) It would be horizontal.
    (e) It would coincide exactly with the dashed curve.

25. Suppose a correlation of $r_k = +0.4825$ is found between the speed, in kilometers per hour, at which a car is driven and the average number of traffic accidents per hour. One statute mile is about 1.609 kilometers. Given this information, what is the approximate correlation r_m between the speed, in statute miles per hour, at which a car is driven and the average number of traffic accidents per hour?
    (a) $r_m = +0.7763$
    (b) $r_m = +0.2999$
    (c) $r_m = +0.4825$
    (d) More information is needed to determine this.
    (e) It can't be defined because the units have changed.

26. If a distribution is bimodal, it means that
    (a) there are two entirely different curves that can represent it
    (b) the data is ambiguous
    (c) there are two different values of the independent variable for which the dependent variable reaches a maximum
    (d) there are two different values of the dependent variable for which the independent variable reaches a minimum
    (e) there are two different but equally valid values for the median

27. Imagine that a correlation of $r_i = -0.500$ is found between the average number of minutes per day people use the Internet and the average number of minutes per day they spend reading books. This logically and rigorously implies
    (a) that Internet usage causes people to spend more time reading books
    (b) that Internet usage causes people to spend less time reading books
    (c) that book reading causes people to spend more time on the Internet
    (d) that book reading causes people to spend less time on the Internet
    (e) none of the above

28. Which of the following methods is best for generating pseudorandom digits?
    (a) Repeatedly spin a wheel calibrated in digits from 0 to 9, with each digit having the same angular space (36° of arc) around the circle.
    (b) Have a child utter digits from 0 to 9 aloud for an indefinite period of time, after being told to say them at random.
    (c) Throw darts at a checker board with squares labeled with digits from 0 to 9, with each square having equal area, there being an equal number of squares for each of the digits 0 through 9.
    (d) Place 10 marbles, one labeled with each digit 0 to 9, in a jar, shake the jar for a minute, then close your eyes and pick a marble out. Repeat this process for each digit needed.
    (e) None of the above methods is any good at all.

29. A regression curve shows
    (a) the extent of the inference for a specific hypothesis
    (b) the general way in which two variables are related
    (c) the probability that the null hypothesis is true
    (d) the probability that a specific alternative hypothesis is true
    (e) nothing significant unless it is a straight line

30. Figure Test 2-4 shows the 95% confidence interval in a normal distribution. This means that

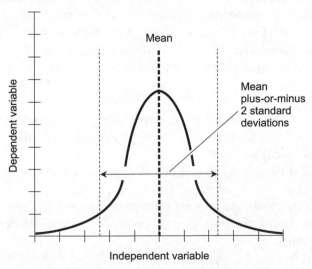

Fig. Test 2-4. Illustration for Part Two Test Questions 30 and 31.

(a) the area under the curve between the light dashed vertical line on
        the left and the light dashed vertical line on the right is equal to
        95% of the total area under the curve
   (b) the area under the curve between the light dashed vertical line on
        the left and the heavy dashed vertical line in the center is equal to
        95% of the total area under the curve
   (c) the area under the curve between the heavy dashed vertical line in
        the center and the light dashed vertical line on the right is equal to
        95% of the total area under the curve
   (d) the total area under the curve is 95%
   (e) none of the above

31. Suppose, in Fig. Test 2-4, the light, vertical dashed lines represent the
     mean plus-or-minus 3 standard deviations, rather than the mean plus-
     or-minus 2 standard deviations. Then the graph would portray
     (a) a 68% confidence interval
     (b) a 95% confidence interval
     (c) a 99.7% confidence interval
     (d) a 100% confidence interval
     (e) a confidence interval that cannot be defined

32. Imagine there has been a heavy snowfall. You are about to use the
     snow blower to remove the snow from your driveway. You say, "It
     will take me exactly 30 minutes to clear the snow from the driveway."
     Your older sister says, "It will take you longer than that." Your
     sister's hypothesis is
     (a) the null hypothesis
     (b) a one-sided alternative hypothesis
     (c) a two-sided alternative hypothesis
     (d) impossible to prove
     (e) valid only if you don't own a snow blower

33. The line of least squares in a scatter plot is always
     (a) a circle
     (b) a parabola
     (c) straight
     (d) horizontal
     (e) vertical

34. The tendency for small events to have dramatic long-term and large-
     scale consequences is called
     (a) correlation magnification
     (b) the Mandelbrot effect

(c) the butterfly effect

(d) regression

(e) standard deviation

35. Which of the following is not a measure of central tendency in a statistical distribution?

(a) The mean.

(b) The median.

(c) The mode.

(d) The dependent variable.

(e) All of the above are measures of central tendency in a statistical distribution.

36. In a statistical distribution, the standard deviation is equal to

(a) the square root of the mean

(b) the square root of the variance

(c) the square root of the median

(d) the square root of the mode

(e) any of the above

37. Random sampling is done in an attempt to

(a) generate a random-number table

(b) get an unbiased cross-section of a population

(c) distort or skew the results of a statistical experiment

(d) ensure that a distribution is normal

(e) define a confidence interval

38. Imagine tossing a pair of identical, unbiased, 6-sided dice. What is the probability that both dice will come up showing 2?

(a) 1 in 2

(b) 1 in 6

(c) 1 in 12

(d) 1 in 36

(e) 1 in 72

39. The standard deviation in a statistical distribution is a measure of

(a) dispersion

(b) central tendency

(c) independent variation

(d) dependent variation

(e) the mode

40. The fact that events often occur in bunches, and the fact that improvement in athletic performance often takes place in spurts rather than steadily and gradually, are explainable according to
    (a) the law of averages
    (b) statistical distributions
    (c) probability theory
    (d) chaos theory
    (e) the theory of the least upper bound

41. A sampling frame is
    (a) a point of view from which a statistical experiment is done
    (b) a range of dependent-variable values
    (c) the same thing as a confidence interval
    (d) a set of items within a population from which a sample is chosen
    (e) a single element in a population

42. Suppose you step on a digital scale that displays your weight in pounds, all the way down to the hundredth of a pound. You are told the scale is accurate to within ±1%. The scale indicates your weight as 120.00 pounds. This means your actual weight could be anywhere between
    (a) 119.99 and 120.01 pounds
    (b) 119.88 and 120.12 pounds
    (c) 118.80 and 121.20 pounds
    (d) 108.00 and 132.00 pounds
    (e) two limits that require more information to figure out

43. Refer to the correlation plot of Fig. Test 2-5. Suppose the dashed line represents the least-squares line for all the solid black points. If a new value is added in the location shown by the gray point P, but no other new values are added, what will happen to the least-squares line?
    (a) It will vanish.
    (b) It will move up from the position shown.
    (c) It will move down from the position shown.
    (d) Its position will not change from that shown.
    (e) More information is needed to answer this question.

44. Refer to the correlation plot of Fig. Test 2-5. Suppose the dashed line represents the least-squares line for all the solid black points. If a new value is added in the location shown by the gray point Q, but no other new values are added, what will happen to the least-squares line?
    (a) It will vanish.

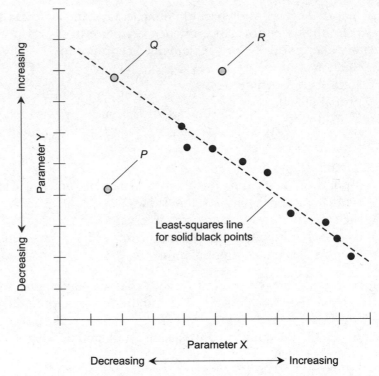

Fig. Test 2-5.   Illustration for Part Two Test Questions 43 through 45.

(b) It will move up from the position shown.
(c) It will move down from the position shown.
(d) Its position will not change from that shown.
(e) More information is needed to answer this question.

45. Refer to the correlation plot of Fig. Test 2-5. Suppose the dashed line represents the least-squares line for all the solid black points. If a new value is added in the location shown by the gray point *R*, but no other new values are added, what will happen to the least-squares line?
(a) It will vanish.
(b) It will move up from the position shown.
(c) It will move down from the position shown.
(d) Its position will not change from that shown.
(e) More information is needed to answer this question.

46. Imagine tossing a pair of identical, unbiased, 6-sided dice. What is the probability that both dice will come up showing the same number of dots?

    (a) 1 in 2
    (b) 1 in 6
    (c) 1 in 12
    (d) 1 in 36
    (e) 1 in 72

47. Suppose a scatter plot shows a strong negative correlation between two variables. How many least-squares lines can exist for this plot?
    (a) None.
    (b) One.
    (c) Two.
    (d) More than two.
    (e) Infinitely many.

48. Refer to Fig. Test 2-6. What is the probability that the randomly chosen point lies within 3 standard deviations of the mean?
    (a) 68%

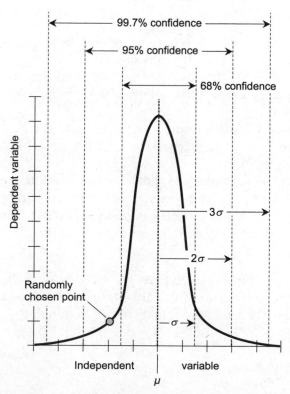

Fig. Test 2-6. Illustration for Part Two Test Question 48.

(b) 95%

(c) 99.7%

(d) Somewhere between 68% and 95%

(e) Somewhere between 95% and 99.7%

49. There are two major ways in which an error can be made when for-mulating hypotheses. One type of error involves rejecting or denying the potential truth of a null hypothesis, and then having the experi-ment show that it's true. The other major type of blunder is to

(a) reject all the alternative hypotheses, and then have the null hypothesis turn out to be true

(b) take a random sample of the population when it's better to take a biased sample

(c) take a population sample that is too large, resulting in an inade-quate cross-sectional representation

(d) accept the null hypothesis and then have the experiment show that it's false

(e) accept all the alternative hypotheses, and have them all turn out true

50. Refer to Fig. Test 2-7. What, if anything, is wrong with this graph?

(a) Nothing is wrong. It's perfectly all right.

(b) The plots for the absolute frequency and the cumulative absolute frequency are labeled wrong, although the axes are labeled cor-rectly.

(c) The axes for the absolute frequency and the cumulative absolute frequency are labeled wrong, although the plots are labeled correctly.

(d) Both the axes and the plots for the absolute frequency and the cumulative absolute frequency are labeled wrong.

(e) Absolute frequency and cumulative absolute frequency can never be plotted together in the same graph, so there's no way to make it correct.

51. Suppose you flip a coin 20 times. What is the probability that the coin will come up "heads" on all 20 flips? Assume the coin is not "weighted," so the probability of it coming up "heads" on any given flip is 50%.

(a) 1 in 20

(b) 1 in 48

(c) 1 in 256

(d) 1 in 512

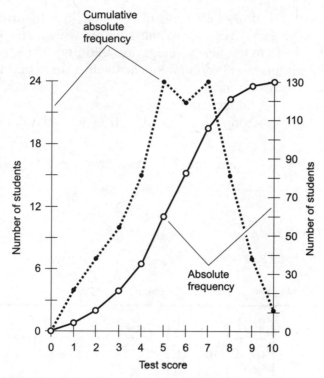

Fig. Test 2-7. Illustration for Part Two Test Question 50.

    (e) None of the above

52. Refer to Table Test 2-3. This shows the results of a hypothetical 10-question test given to a large class of students. What score represents the mode?
    (a) 5
    (b) 6
    (c) 45
    (d) 59
    (e) 262

Table Test 2-3 Table for Part Two Test Questions 52 and 54.

Test score	0	1	2	3	4	5	6	7	8	9	10
Absolute frequency	0	6	10	14	22	45	59	37	33	21	15

53. Table Test 2-4 shows the results of the same hypothetical test as that portrayed in Table Test 2-3. What is wrong with Table Test 2-4?
(a) The absolute frequency values don't add up correctly.
(b) The cumulative absolute frequency values don't add up correctly.
(c) There exists no mean.
(d) There exists no median.
(e) Nothing is wrong with this table. It is perfectly plausible.

54. In Tables Test 2-3 and Test 2-4, what score represents the median?
(a) 5
(b) 6
(c) 45
(d) 59
(e) 262

Table Test 2-4   Table for Part Two Test Questions 53 and 54.

Test score	0	1	2	3	4	5	6	7	8	9	10
Absolute frequency	0	6	10	14	22	45	59	37	33	21	15
Cumulative absolute frequency	0	6	16	30	52	97	156	193	226	247	262

55. In a vertical bar graph, the independent variable is normally portrayed
(a) along the horizontal axis
(b) along the vertical axis
(c) as the heights of various rectangles or bars
(d) as the widths of various rectangles or bars
(e) as a smooth curve

56. In a horizontal bar graph, the independent variable is normally portrayed
(a) as the heights of various rectangles or bars
(b) as the widths of various rectangles or bars
(c) along the horizontal axis
(d) along the vertical axis

(e) as a smooth curve

57. Fill in the following sentence to make it the most accurate: "It's possible to _____ express the correlation between two variables if one or both of them cannot be quantified."
(a) precisely
(b) qualitatively
(c) inversely
(d) numerically
(e) logically

58. Once in a while, you'll see a scatter plot in which almost all of the points lie near a straight line, but there are a few points that are far away from the main group. Stray points of this sort are called
(a) wingers
(b) uncorrelaters
(c) nonconformers
(d) outliers
(e) scatterers

59. What is the maximum possible number of hypotheses in a scenario?
(a) One: the null hypothesis.
(b) Two: the null hypothesis and the alternative hypothesis.
(c) Three: the null hypothesis, a one-sided alternative hypothesis, and a two-sided alternative hypothesis.
(d) Four: the null hypothesis, two one-sided alternative hypotheses, and a two-sided alternative hypothesis.
(e) It is impossible to answer this question without knowing more about the scenario.

60. Suppose you want to determine the average (mean) time that Canadian postal workers get out of bed in the morning. You conduct a survey of 50 postal workers in various Canadian cities and towns, and find that the average time they arise is 6:18 A.M. This time is
(a) a population of the mean
(b) an estimate of the mean
(c) precisely equal to the mean
(d) a sampling frame of the mean
(e) a meaningless statistic

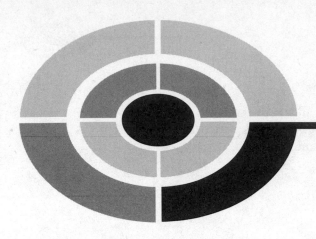

Final Exam

Do not refer to the text when taking this test. You may draw diagrams or use a calculator if necessary. A good score is at least 75 correct. Answers are in the back of the book. It's best to have a friend check your score the first time, so you won't memorize the answers if you want to take the test again.

1. The term *empirical probability* refers to
   (a) the probability of something going wrong in an experiment
   (b) the probability of a certain outcome, based on experience or observation
   (c) the probability that an experiment accurately represents real life
   (d) the probability that a certain event took place sometime in the past
   (e) the probability that a certain event is taking place right now

2. In a statistical distribution, the variance is
   (a) a measure of the extent to which the values are spread out
   (b) a measure of the correlation between the variables
   (c) an expression of the difference between the mean and the median
   (d) the distance of the mode from the center of the distribution

(e) a measure of experimental error

3. Suppose q and r are both positive integers. Imagine a set of q items taken r at a time in a specific order. The possible number of permutations in this situation is symbolized $_qP_r$ and can be calculated as follows:

$$_qP_r = q! \ / \ (q - r)!$$

According to this formula, what is the number of possible permutations of 5 objects taken 3 at a time?
(a) 10
(b) 20
(c) 60
(d) 120
(e) More information is needed to answer this.

4. Table Exam-1 shows the results of a hypothetical experiment in which a 6-sided die is tossed 6000 times. What, if anything, is wrong with this table?
(a) There is nothing wrong with the table; it is entirely plausible.
(b) The data for the absolute frequency and the cumulative absolute frequency are in the wrong columns.
(c) There is no way a coincidence like this could ever occur.
(d) The absolute frequency values don't add up right.
(e) The cumulative absolute frequency values don't add up right.

Table Exam-1   Table for Final Exam Questions 4 and 5.

Face of die	Absolute frequency	Cumulative absolute frequency
1	1000	1000
2	1000	2000
3	1000	3000
4	1000	4000
5	1000	5000
6	1000	6000

5. If a 6-sided die is tossed 6000 times and the results turn out like those shown in Table Exam-1, it is reasonable to conclude that
   (a) the die is biased or "weighted"
   (b) the die is not biased or "weighted"
   (c) the sampling frame is too large
   (d) the mean is equal to 0
   (e) the mode is equal to 0

6. Suppose q and r are both positive integers. Imagine a set of q items taken r at a time in no particular order. The possible number of combinations in this situation is symbolized $_qC_r$ and can be calculated as follows:

$$_qC_r = q! \,/\, [r!(q - r)!]$$

   According to this formula, what is the number of possible combinations of 5 objects taken 3 at a time?
   (a) 10
   (b) 20
   (c) 60
   (d) 120
   (e) More information is needed to answer this.

7. What is the arithmetic mean, or average, of a group of numbers represented by the variables t, u, v, and w?
   (a) $(t + u + v + w) \,/\, 4$
   (b) $tuvw \,/\, 4$
   (c) $(t + u + v + w) \,/\, tuvw$
   (d) $tuvw \,/\, (t + u + v + w)$
   (e) There is no way to express it without knowing actual number values.

8. Figure Exam-1 is an example of
   (a) a histogram
   (b) a pie graph
   (c) a bar graph
   (d) a pizza graph
   (e) a tablet graph

9. What is the angle at the apex (or central "point") of the wedge corresponding to 34.0% in Fig. Exam-1, to the nearest degree?
   (a) 34°
   (b) 56°
   (c) 122°

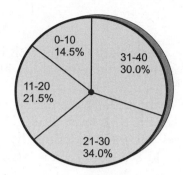

Fig. Exam-1.   Illustration for Final Exam Questions 8 and 9.

(d) 244°
(e) This can't be answered without more information.

10. Imagine the set of all people in the world who read and write English. Call this set *E*. Now imagine the set of all people in the world. Call this set *W*. What can we say about the relationship between these two sets?
(a) *E* is a subset of *W*.
(b) *E* is an element of *W*.
(c) *W* is a subset of *E*.
(d) *W* is an element of *E*.
(e) None of the above

11. Figure Exam-2 shows the letter-grade results of a hypothetical test given to a large group of students. This illustration is an example of
(a) a horizontal bar graph
(b) a fixed-width histogram
(c) a variable-width histogram
(d) a pie graph

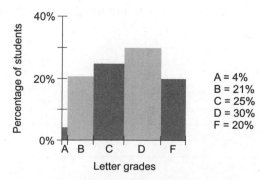

Fig. Exam-2.   Illustration for Final Exam Questions 11 and 12.

(e) a nomograph

12. What, if anything, is technically wrong with Fig. Exam-2?
(a) The rectangles are not the correct heights to correspond to the percentages shown at the right.
(b) The rectangles' widths don't reasonably represent the proportion of students receiving each letter grade.
(c) The percentages don't add up correctly.
(d) No teacher would ever give such a large proportion of students low grades.
(e) Nothing is technically wrong with Fig. Exam-2.

13. Fill in the blank to make the following sentence true: "The _____ of a particular outcome is the number of times it occurs within a specific sample of a population."
(a) frequency
(b) variability
(c) variance
(d) standard deviation
(e) distribution

14. Table Exam-2 illustrates the results of a test given to a group of students. What, if anything, is wrong or implausible with this table?
(a) The percentages don't add up right.
(b) The entries in the letter-grade column are listed upside-down.
(c) The absolute frequency numbers don't correspond to the correct percentages.
(d) It's impossible to tell how many students there are.

Table Exam-2   Table for Final Exam Question 14.

Letter grade	Range of scores	Absolute frequency	Percentage of scores
A	38–40	60	30%
B	32–37	40	20%
C	25–31	50	25%
D	19–24	42	21%
F	0–18	8	4%

(e) Nothing is wrong or implausible with this table.

15. Suppose you are told to calculate the probability of a certain event, and it turns out as $P = 0.78544$. What is this expressed as a percentage $P_\%$, rounded off to the nearest percentage point?
    (a) 0.78%
    (b) 0.79%
    (c) 78%
    (d) 79%
    (e) More information is needed to answer this.

16. Fill in the blank to make the following sentence correct: "A specific, well-defined characteristic of a population is called a _____ of that population."
    (a) variable
    (b) mean
    (c) parameter
    (d) distribution
    (e) central limit

17. A continuous variable can attain any value between certain limits, but a discrete variable can attain
    (a) only one value
    (b) only the values of the limits themselves
    (c) only specific values
    (d) no values at all
    (e) only values that produce a straight-line graph

18. The set of all possible outcomes in an experiment is called
    (a) the variance
    (b) the sample space
    (c) the area under the curve
    (d) the normal distribution
    (e) the total probability

19. An error in an experiment can be caused by
    (a) imprecise visual interpolation of an instrument reading
    (b) proper tallying of the data
    (c) a machine that is working properly
    (d) rendering of the data in the form of a graph
    (e) any of the above

20. In a normal distribution, the Z score is a quantitative measure of the position of a particular element with respect to

(a) the mean
(b) the median
(c) the mode
(d) the standard deviation
(e) the variance

21. Suppose you take a test with 100 questions, and you get a score of 77. Which, if any, of the following statements (a), (b), (c), or (d) can you make with certainty?
    (a) You scored in the 77th percentile.
    (b) You scored in the 78th percentile.
    (c) You scored in the 7th decile.
    (d) You scored in the 3rd quartile.
    (e) You can't make any of the above statements (a), (b), (c), or (d) with certainty.

22. Suppose you are analyzing a normal distribution. You have good estimates of the mean and standard deviation. Which of the following can be determined on the basis of this information?
    (a) The 50% confidence interval.
    (b) The 60% confidence interval.
    (c) The 70% confidence interval.
    (d) The 80% confidence interval.
    (e) All of the above (a), (b), (c), and (d) can be determined on the basis of this information.

23. Suppose you want to determine the percentage of people in your county who swim more than 1000 meters a day. Which of the following samples should you expect would be the best (least biased)?
    (a) All the people in the county over age 75.
    (b) All the hospital patients in the county.
    (c) All the people in the county who were born in August.
    (d) All the high-school students in the county.
    (e) All the people in the county who weigh less than 100 pounds.

24. As the width of the confidence interval in a normal distribution increases, the probability of a randomly selected element falling within that interval
    (a) approaches 68%
    (b) approaches 50%
    (c) increases
    (d) decreases
    (e) does not change

25. Suppose you buy a certificate of deposit (CD) at your bank. Once you've bought it, you leave it alone, allowing it to mature automatically and earn interest indefinitely. The value of this CD, in dollars, can be plotted against time, and the result is a function. Assuming the CD never loses value (the interest never becomes negative), this function is
    (a) nonincreasing
    (b) nondecreasing
    (c) linear
    (d) point-to-point
    (e) a histogram

26. An exclamation mark (!) following a positive whole number represents
    (a) the factorial
    (b) the mean
    (c) the median
    (d) the mode
    (e) the cumulative frequency

27. Refer to Fig. Exam-3. In this graph, the stock price is the
    (a) independent variable
    (b) dependent variable
    (c) mean
    (d) standard deviation
    (e) variance

28. In Fig. Exam-3, how might a statistician attempt to fill in the gap in the data, even if no actual data is available?

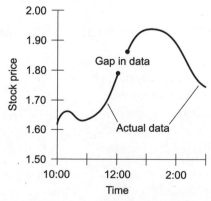

Fig. Exam-3. Illustration for Final Exam Questions 27 and 28.

(a) By means of linear extrapolation.
(b) By finding the least-squares line.
(c) By constructing a histogram.
(d) By means of linear interpolation.
(e) By invoking the law of averages.

29. The bias in an experiment can be minimized by
    (a) minimizing the number of samples
    (b) proper choice of the sampling frame
    (c) choosing a small population
    (d) using properly calibrated measuring equipment
    (e) all of the above

30. Which of the following is implied by the Central Limit Theorem?
    (a) As the size of a population increases beyond 30, the mean approaches the median.
    (b) As the size of a population increases beyond 30, the median approaches the mode.
    (c) If the sample size is 30 or more, then the sampling distribution of means is essentially a normal distribution.
    (d) In a bimodal distribution, the mean is always the same as the median if the population is 30 or more.
    (e) In a bimodal distribution, the mean approaches the median as the size of the population increases beyond 30.

31. Which of the following statements is true?
    (a) Two sets can be proper subsets of each other.
    (b) Two sets can be elements of each other.
    (c) The union of two sets is always at least as big as the smaller one.
    (d) The intersection of two sets is always at least as big as the smaller one.
    (e) The intersection of two sets can never be the same as their union.

32. In Fig. Exam-4, the sampling frame
    (a) coincides with the sample
    (b) is a subset of the sample
    (c) coincides with the population
    (d) is a subset of the population
    (e) is too small

33. When we draw specific conclusions on the basis of data and hypotheses, we are performing
    (a) linear interpolation

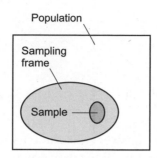

Fig. Exam-4. Illustration for Final Exam Question 32.

(b) curve fitting
(c) statistical inference
(d) least-squares analysis
(e) correlation determination

34. Percentiles divide a large data set into intervals, each interval containing 1% of the elements in the set. The percentile points represent the boundaries where the intervals meet. How many possible percentile points are there?
(a) 4
(b) 24
(c) 25
(d) 99
(e) 100

35. Imagine that 2020 people take a 100-question test. Suppose 20 students get 100 correct answers, 20 students get 99 correct, 20 students get 98 correct, and so on, all the way down to 20 students getting none correct. For every possible test score, 20 students get that score. What is the mean score, accurate to two decimal places?
(a) 49.55
(b) 50.00
(c) 50.45
(d) 0
(e) It cannot be defined.

36. Imagine that 2020 people take a 100-question test. Suppose 20 students get 100 correct answers, 20 students get 99 correct, 20 students get 98 correct, and so on, all the way down to 20 students getting none correct. For every possible test score, 20 students get that score. What is the mode score, accurate to two decimal places?

(a) 49.55
(b) 50.00
(c) 50.45
(d) 0
(e) It cannot be defined.

37. Imagine that 2020 people take a 100-question test. Suppose 20 students get 100 correct answers, 20 students get 99 correct, 20 students get 98 correct, and so on, all the way down to 20 students getting none correct. For every possible test score, 20 students get that score. What is the median score, accurate to two decimal places?
(a) 49.55
(b) 50.00
(c) 50.45
(d) 0
(e) It cannot be defined.

38. Let x be a discrete random variable that can attain n possible values, all equally likely. Suppose an outcome H results from exactly m different values of x, where $m \leq n$. Then m/n represents
(a) the empirical probability that H will result from any given value of x
(b) the mathematical probability that H will result from any given value of x
(c) the discrete probability that H will result from any given value of x
(d) the continuous probability that H will result from any given value of x
(e) none of the above

39. Suppose you take a standardized test. The teacher says you scored in the 4th percentile. You ask, "What is the highest-scoring percentile?" The teacher replies, "The 1st." Based on this, you can be certain that you scored
(a) in the highest 1% of the class
(b) in the highest 25% of the class
(c) in the second-highest 25% of the class
(d) in the second-lowest 25% of the class
(e) in the lowest 25% of the class

40. Imagine that you conduct an experiment with a penny that is unbiased, so the probability is 50% that it will come up "heads"

and 50% that it will come up "tails" on any given toss. If you toss this coin n times, where n is a natural number and is at least 1, the probability P_n that the coin will come up "heads" on every toss is

(a) $P_n = 1/n$
(b) $P_n = 2^n$
(c) $P_n = 1/(2^n)$
(d) $P_n = n/2$
(e) none of the above

41. Suppose you plot the formula derived in the previous question as points on a graph, and then connect the points by curve fitting. Which of the curves in Fig. Exam-5 is the best representation of the result?

(a) A
(b) B
(c) C

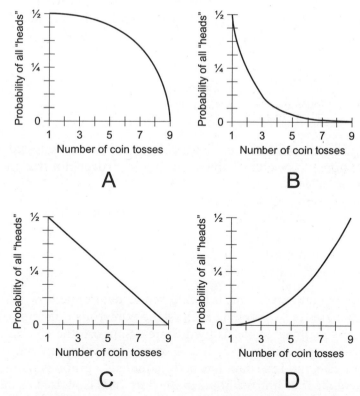

Fig. Exam-5. Illustration for Final Exam Questions 40 and 41.

(d) D

(e) None of them.

42. A variable whose value cannot be predicted in any given instance is called
(a) a random variable
(b) a discrete variable
(c) a continuous variable
(d) a dependent variable
(e) an independent variable

43. Time is often portrayed in graphs as the
(a) mode
(b) standard deviation
(c) variance
(d) independent variable
(e) line of least squares

44. Inaccuracy can be introduced into an experiment when a person reads an analog instrument such as a conventional "pointer style" meter, even if the meter is working perfectly, because of
(a) interpolation error
(b) a sampling frame that is too small
(c) failure to choose a random sample
(d) the choice of a biased sample
(e) confusion of independent and dependent variables

45. When a hypothesis is made with the intent that it be tested, ultimately (we hope) to be either shown true or else proven not true, that hypothesis is called
(a) a propositional hypothesis
(b) a sample hypothesis
(c) a subjective hypothesis
(d) a null hypothesis
(e) a median hypothesis

46. Suppose an experiment is conducted to determine how many human-hours are spent watching each of the numerous different satellite television channels in the United States. Suppose the channels are arbitrarily numbered 1, 2, 3, 4, and so on. After the experiment has been completed and the data has been gathered, a graph is plotted in which the number of human-hours in the year 2003 is plotted as a function of the channel number. The channel numbers constitute

(a) a continuous variable
(b) a discrete variable
(c) a dependent variable
(d) a horizontal variable
(e) a frequency variable

47. Table Exam-3 shows an example of
   (a) ordered sampling
   (b) continuous sampling
   (c) sampling with adjustment
   (d) sampling with replacement
   (e) sampling with correlation

48. Which of the following correlation figures (a), (b), (c), or (d) is implausible?
   (a) $r = +0.5$
   (b) $r = +25\%$
   (c) $r = -75\%$
   (d) $r = -5$
   (e) All of the above (a), (b), (c), and (d) are plausible.

49. Suppose you want to figure out the quantitative effect that the consumption of fat has on people's cholesterol levels. You ask some people how much fat they eat, and you measure their cholesterol levels. The population for your experiment is the set of all people in the world. You choose a sampling frame that consists of 10 people from each officially recognized country in the world. Which, if any, of the following (a), (b), or (c) points to a potential flaw in this scheme?
   (a) People cannot in general accurately tell how much fat they eat.
   (b) The sampling frame is extremely small.
   (c) Factors other than fat consumption can affect cholesterol levels.
   (d) None of the above (a), (b), or (c) points to a flaw in the scheme.
   (e) All of the above (a), (b), and (c) point to flaws in the scheme.

50. Fill in the blank to make the following sentence true: "If the measurement unit of either variable in a graph is changed in size but still refers to the same phenomenon (for example, miles to kilometers or degrees Celsius to degrees Fahrenheit), the plot may be distorted vertically or horizontally, but the _____, if any, between the two variables is not affected."
   (a) standard deviation
   (b) variance

Table Exam-3 Table for Final Exam Question 47.

Before sampling	Elements	After sampling
{αβγδεζηθκλμνξοπ}	λ	{αβγδεζηθκλμνξοπ}
{αβγδεζηθκλμνξοπ}	ξ	{αβγδεζηθκλμνξοπ}
{αβγδεζηθκλμνξοπ}	π	{αβγδεζηθκλμνξοπ}
{αβγδεζηθκλμνξοπ}	η	{αβγδεζηθκλμνξοπ}
{αβγδεζηθκλμνξοπ}	μ	{αβγδεζηθκλμνξοπ}
{αβγδεζηθκλμνξοπ}	ε	{αβγδεζηθκλμνξοπ}
{αβγδεζηθκλμνξοπ}	δ	{αβγδεζηθκλμνξοπ}
{αβγδεζηθκλμνξοπ}	κ	{αβγδεζηθκλμνξοπ}
{αβγδεζηθκλμνξοπ}	β	{αβγδεζηθκλμνξοπ}
{αβγδεζηθκλμνξοπ}	α	{αβγδεζηθκλμνξοπ}
{αβγδεζηθκλμνξοπ}	θ	{αβγδεζηθκλμνξοπ}
{αβγδεζηθκλμνξοπ}	ν	{αβγδεζηθκλμνξοπ}
{αβγδεζηθκλμνξοπ}	ζ	{αβγδεζηθκλμνξοπ}
{αβγδεζηθκλμνξοπ}	γ	{αβγδεζηθκλμνξοπ}
{αβγδεζηθκλμνξοπ}	ο	{αβγδεζηθκλμνξοπ}
↓	↓	↓

    (c) correlation

    (d) mode

    (e) median

51. Imagine two sets that have no elements in common. These sets are said to be

(a) null-coincident
(b) minimally coincident
(c) intersecting
(d) element-free
(e) disjoint

52. Figure Exam-6 illustrates two normal distributions, represented by curves X and Y. Which, if any, of the following statements (a), (b), (c), or (d) is false?
    (a) The standard deviation of the distribution represented by curve X differs from the standard deviation of the distribution represented by curve Y.
    (b) The variance of the distribution represented by curve X differs from the variance of the distribution represented by curve Y.
    (c) The coefficient of variation of the distribution represented by curve X differs from the coefficient of variation of the distribution represented by curve Y.
    (d) The mean of the distribution represented by curve X differs from the mean of the distribution represented by curve Y.
    (e) All of the above statements (a), (b), (c), and (d) are true.

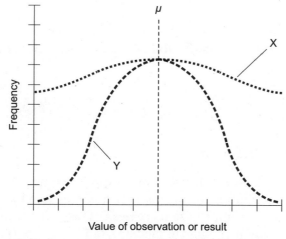

Fig. Exam-6. Illustration for Final Exam Question 52.

53. What is the smallest number of elements a set can have?
    (a) Less than 0
    (b) 0
    (c) 1
    (d) More than 1

(e) Infinitely many

54. What, if anything, is wrong or implausible with Fig. Exam-7?
(a) The sampling frame should be, but is not, a subset of the population.
(b) The sample should be, but is not, a subset of the sampling frame.
(c) The population should be, but is not, a subset of the sampling
frame.
(d) The sampling frame should be, but is not, a subset of the sample.
(e) Nothing is wrong or implausible with Fig. Exam-7.

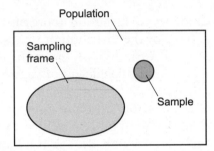

Fig. Exam-7.   Illustration for Final Exam Question 54.

55. Suppose the correlation between two variables is as weak as it can
possibly be. Which of the following expresses the correlation, r, in
quantitative form?
(a) $r = -100$
(b) $r = -1$
(c) $r = -100\%$
(d) $r = -1\%$
(e) None of the above.

56. Consider a statistical distribution. Suppose someone tells you, "The
sampling distribution of means gets less and less like a normal distribution as the sample size increases." This statement
(a) is true only if the mean, median, and mode are all the same
(b) is true only for infinite populations
(c) is true only for large populations
(d) is true only for small populations
(e) is patently false

57. The decimal expansion of the square root of 10
(a) is a relation but not a function

(b) is a function but not a relation
(c) cannot be written out in its entirety
(d) is an integer
(e) is not a real number

58. Imagine a scatter plot in which almost all of the points lie near a straight line, but there are a few points that are far away from the main group. Stray points of this sort are called
(a) nonlinear points
(b) points of greatest squares
(c) points of least squares
(d) error points
(e) none of the above

59. Figure Exam-8 is an example of
(a) a scatter plot
(b) a point-to-point graph
(c) linear interpolation
(d) a least-squares line

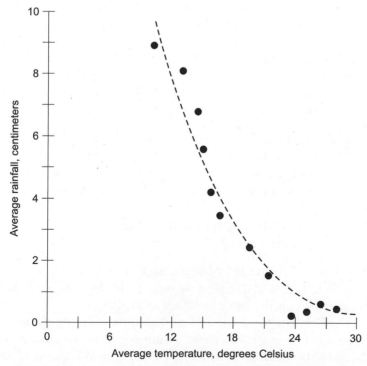

Fig. Exam-8. Illustration for Final Exam Questions 59 and 60.

(e) a quadratic equation

60. The dashed curve in Fig. Exam-8 represents
    (a) a least-squares line
    (b) a point-to-point graph
    (c) linear interpolation
    (d) a regression curve
    (e) a quadratic equation

61. Consider the set of all people in the state of Arizona who smoke cigarettes. Call this set A_s. Suppose John lives in Arizona and smokes cigarettes. John is
    (a) a sampling frame with respect to A_s
    (b) a subset of A_s
    (c) a proper subset of A_s
    (d) an element of A_s
    (e) a population frame of A_s

62. One of the major ways in which an error can be made when formulating hypotheses is to assume that the null hypothesis is false, and then
    (a) have the experiment confirm that it is false
    (b) have the experiment show that it is true
    (c) have the experiment show that one alternative hypothesis is true
    (d) have the experiment show that two of the alternative hypotheses are true
    (e) have the experiment show that all of the alternative hypotheses are true

63. Pseudorandom numbers
    (a) can be generated by having a child chatter off digits out loud
    (b) are theoretically random
    (c) are equivalent to theoretically random numbers for practical purposes
    (d) can be generated by spinning a wheel
    (e) have a correlation of −1

64. Table Exam-4 shows temperature and rainfall data for a hypothetical town. What, if anything, is wrong with this table? If anything is wrong, how can the table be made correct?
    (a) The data in the left-hand and middle columns is entered incorrectly. The entries in these columns should be transposed.
    (b) The data in the left-hand and right-hand columns is entered incorrectly. The entries in these columns should be transposed.

Table Exam-4   Table for Final Exam Questions 64 and 65.

Month	Average temperature, degrees Celsius	Average monthly rainfall, centimeters	Cumulative rainfall for year, centimeters
January	2.1	0.4	0.4
February	3.0	0.6	0.2
March	9.2	0.9	0.3
April	15.2	3.7	2.8
May	20.4	7.9	4.2
June	24.9	14.2	6.3
July	28.9	21.5	7.3
August	27.7	30.0	8.5
September	25.0	37.7	7.7
October	18.8	41.3	3.6
November	10.6	43.0	1.7
December	5.3	43.5	0.5

    (c) The data in the middle and right-hand columns is entered incorrectly. The entries in these columns should be transposed.

    (d) The data in the left-hand column does not add up correctly. The addition errors should be corrected.

    (e) Nothing is wrong with Table Exam-4.

65. Assuming that Table Exam-4 is correct as shown, or that it is corrected if it contains errors, what can be said about the correlation between the average monthly temperature and the average monthly rainfall for this hypothetical town?

    (a) There is no correlation.

(b) It is negative.

(c) It is positive.

(d) It is a null hypothesis.

(e) Nothing. It is impossible to tell if there is correlation or not.

66. Suppose a computer is programmed to find the correlation between two variables, and the correlation is found to be 0. The computer is then programmed to produce a scatter plot and identify the least-squares line (if there is one) for the points in the plot. The computer will show us that the least-squares line

(a) ramps downward as you move toward the right

(b) ramps upward as you move toward the right

(c) is not straight, but is a parabola

(d) is not straight, but is a hyperbola

(e) does not exist

67. Suppose a computer is programmed to find the correlation between two variables, and the correlation is found to be 0. The computer is then programmed to produce a scatter plot. The points in the plot will most likely

(a) be spread out all over the graph

(b) lie along a line or curve that ramps upward as you move toward the right

(c) lie along a line or curve that ramps downward as you move toward the right

(d) lie along a parabola

(e) lie along a hyperbola

68. Imagine that tornadoes occur in a given county on the average of one event every 3 years. Suppose 10 years go by and there are no tornadoes in that county. Then, in the next year, there are 3 tornadoes. This should not surprise us because

(a) bunching-up of events simply happens in nature from time to time

(b) the law of averages forced it

(c) tornadoes were "due" in the county

(d) nature built up a "tornado deficit" in the county that had to be "paid off"

(e) the butterfly effect forced it to happen

69. Two outcomes are mutually exclusive if and only if

(a) they always occur in every situation

(b) they have some of their elements in common

(c) they have all their elements in common

    (d) they have no elements in common

    (e) they never occur in any situation

70. When sampling is done with replacement in a finite set:
    (a) the size of the set does not change
    (b) the size of the set increases
    (c) the size of the set decreases
    (d) the size of the set becomes negative
    (e) none of the above

71. The butterfly effect is responsible for the fact that
    (a) variables are always correlated, even when it seems that they are not
    (b) cause–effect often works in the opposite way from what we would expect
    (c) localized, small events can have widespread, large-scale consequences
    (d) things always happen in bunches
    (e) if there is an upper bound, then there is a least upper bound

72. A sequence of digits from the set {0, 1, 2, 3, 4, 5, 6, 7, 8, 9} can be considered random
    (a) if and only if they can be written down as a nonterminating, repeating sequence
    (b) if and only if they coincide with the digits in the decimal expansion of some specific irrational number such as the square root of 2
    (c) if and only if, given any digit in the sequence, the next one is a function of it
    (d) if and only if, given any digit in the sequence, there exists no way to predict the next one
    (e) if and only if they are generated by a calculator or computer

73. Suppose the distance between the dashed line and each of the points in Fig. Exam-9 is measured, producing a set of distance numbers. These numbers are squared and then the squares are added up, getting a final sum D. The dashed line in Fig. Exam-9 represents the line for which D is minimized for the set of points shown. This means that the dashed line is the
    (a) mean correlation line
    (b) interpolation line
    (c) least-squares line
    (d) median line

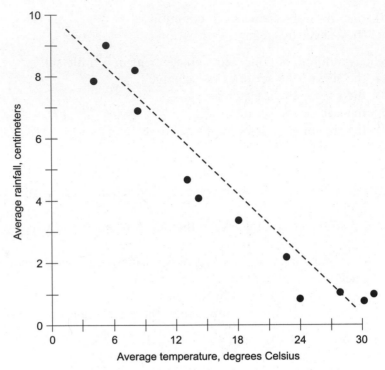

Fig. Exam-9. Illustration for Final Exam Question 73.

(e) function line

74. In a uniform distribution, the value of the function is
(a) peaked at one point
(b) always equal to 0
(c) constant
(d) increasing
(e) decreasing

75. If either scale in a graph of a function has a span that is not large
enough, then
(a) it doesn't really matter, although it might look bad
(b) the dependent and independent variables might be confused with
each other
(c) there won't be enough room to illustrate the whole function, or
the part we're most interested in
(d) space in the graph will be wasted
(e) the graph will appear cluttered

76. In a scatter plot, the correlation is defined by
    (a) the nearness of the points to a particular circle
    (b) the nearness of the points to a particular straight line
    (c) the nearness of the points to the horizontal axis
    (d) the nearness of the points to the vertical axis
    (e) the number of points overall

77. Which of the following situations has probability of 0 in the real world
    – that is, it never occurs?
    (a) The mean, median, and mode in a distribution can all be the same.
    (b) The correlation between two variables can be less than 0.
    (c) Two events can be caused by a third event.
    (d) A minuscule event can have huge consequences.
    (e) None of the above situations has probability 0 in the real world.
        That is to say, any of them can occur.

78. Suppose you are about to relocate to New York City. You wonder
    what proportion of the people there eat sushi at least once a week.
    You assume that the figure is 20%. I think the proportion is higher
    than that, and I will try to prove you wrong. My contention is an
    example of
    (a) a null hypothesis
    (b) an alternative hypothesis
    (c) a standard deviation
    (d) weak correlation
    (e) strong correlation

79. Quartile points break a data set up into intervals, each interval con-
    taining approximately _____ of the elements in the set.
    (a) 1/2
    (b) 1/3
    (c) 1/4
    (d) 1/10
    (e) 1/100

80. Which, if any, of the following (a), (b), (c), or (d) is an example of
    statistical inference?
    (a) An alternative hypothesis.
    (b) A null hypothesis.
    (c) A significance test.
    (d) A confidence interval.
    (e) None of the above (a), (b), (c), or (d) is an example of statistical
        inference.

81. Suppose we want to see how income correlates with age, and we have the financial records of 500 people (which we obtained with their permission). The clearest way to graphically illustrate this correlation, if there is any, is to put the data in the form of
    (a) a table
    (b) a scatter plot
    (c) a null graph
    (d) a pie graph
    (e) a bar graph

82. Sampling is a process of
    (a) analyzing data
    (b) gathering data
    (c) determining correlation
    (d) determining cause-and-effect
    (e) graphing functions

83. What is the real-number solution set for $x^2 - 7x + 12 = 0$?
    (a) $\{-7, 12\}$
    (b) $\{7, -12\}$
    (c) $\{3, 4\}$
    (d) $\{-3, -4\}$
    (e) There are no real-number solutions to this equation.

84. Refer to Fig. Exam-10. The curve is "bell-shaped" and is symmetrical on the left-hand and right-hand sides of the heavy, vertical dashed line labeled $x = \mu$. This is a classical illustration of
    (a) a uniform distribution
    (b) a bimodal distribution
    (c) a discrete distribution
    (d) a normal distribution
    (e) a discontinuous distribution

85. In Fig. Exam-10, the symbol μ is meant to represent
    (a) the mean
    (b) the median
    (c) the mode
    (d) the variance
    (e) the standard deviation

86. In Fig. Exam-10, the symbol σ is meant to represent
    (a) the mean
    (b) the median

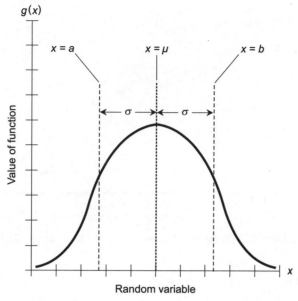

Fig. Exam-10. Illustration for Final Exam Questions 84 through 86.

(c) the mode
(d) the variance
(e) the standard deviation

87. Figure Exam-11 is an example of
(a) a scatter plot
(b) a correlation graph
(c) linear interpolation
(d) curve fitting
(e) a bar graph

88. The numbers to the right of the shaded rectangles in Fig. Exam-11 are there
(a) to clarify the values represented by the rectangles
(b) to make the graph appear less cluttered
(c) to make the graph look more sophisticated
(d) to provide data for making other graphs
(e) for no useful reason

89. Figure Exam-12 shows the results of a 100-question test given to a large group of students. The boundary points in this nomograph define the
(a) quartiles

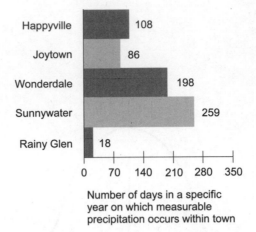

Fig. Exam-11. Illustration for Final Exam Questions 87 and 88.

(b) deciles
(c) percentiles
(d) equal intervals
(e) none of the above

90. What, if anything, is wrong or implausible with the nomograph in Fig. Exam-12?
(a) The boundary points should be equally spaced.
(b) There should be 10 points, not 9.
(c) There should be points representing scores of 0 and 100.
(d) There should be a point representing a score of 50.
(e) Nothing is wrong or implausible with the nomograph in Fig. Exam-12.

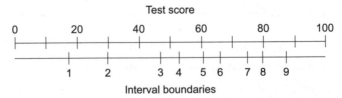

Fig. Exam-12. Illustration for Final Exam Questions 89 and 90.

91. If no data are available, statisticians must collect it themselves. Data collected directly by the statisticians who analyze it is known as
(a) independent source data
(b) dependent source data

(c) primary source data

(d) secondary source data

(e) none of the above

92. As the size of an experimental sample set increases:
    (a) the size of the population increases
    (b) the size of the population decreases
    (c) the estimate of the mean can be done with less and less accuracy
    (d) the estimate of the mean can be done with more and more accuracy
    (e) the standard deviation approaches 0

93. Which of the following statements (a), (b), (c), or (d) is true?
    (a) The values in a bar graph do not necessarily have to add up to 100%.
    (b) Some functions are relations.
    (c) Zero correlation is indicated by widely scattered points on a graph.
    (d) A histogram is a specialized bar graph.
    (e) All of the above statements (a), (b), (c), and (d) are true.

94. Suppose it's autumn in Minnesota, and you predict that it will be an average winter temperature-wise, based on historical data. This is a null hypothesis. Your uncle Jim thinks it will be a colder winter than average. Your sister Susan thinks it will be either warmer or colder than average, but not average. Uncle Jim's prediction is an example of
    (a) a one-sided alternative hypothesis
    (b) a two-sided alternative hypothesis
    (c) a positive hypothesis
    (d) a negative hypothesis
    (e) an off-center hypothesis

95. Suppose it's autumn in Minnesota, and you predict that it will be an average winter temperature-wise, based on historical data. This is a null hypothesis. Your uncle Jim thinks it will be a colder winter than average. Your sister Susan thinks it will be either warmer or colder than average, but not average. Susan's prediction is an example of
    (a) a one-sided alternative hypothesis
    (b) a two-sided alternative hypothesis
    (c) a positive hypothesis
    (d) a negative hypothesis
    (e) an off-center hypothesis

96. Consider the following process for limiting the length of a number to three decimal places:

$$35.78790178$$
$$35.7879017$$
$$35.787901$$
$$35.78790$$
$$35.7879$$
$$35.787$$

The steps in this process are examples of
(a) rounding
(b) variance
(c) normalization
(d) truncation
(e) standard deviation

97. As the number of events in an experiment increases, the average value of the outcome approaches the theoretical mean. This is a statement of
(a) the law of least squares
(b) the Central Limit Theorem
(c) the law of large numbers
(d) the Regression Theorem
(e) the butterfly effect

98. In the plot of Fig. Exam-13, the correlation between phenomenon X and phenomenon Y appears to be
(a) positive
(b) negative
(c) zero
(d) linear
(e) undefined

99. With respect to the plot shown by Fig. Exam-13, which of the following scenarios (a), (b), (c), or (d) is plausible?
(a) Changes in the frequency, intensity, or amount of X cause changes in the frequency, intensity, or amount of Y.
(b) Changes in the frequency, intensity, or amount of Y cause changes in the frequency, intensity, or amount of X.
(c) Changes in the frequency, intensity, or amount of some third factor, Z, cause changes in the frequencies, intensities, and amounts of both X and Y.

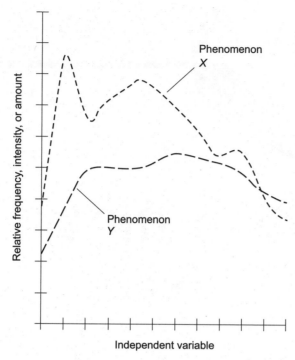

Fig. Exam-13. Illustration for Final Exam Questions 98 and 99.

(d) There is no cause–effect relationship between X and Y whatsoever.

(e) Any of the above scenarios (a), (b), (c), or (d) is plausible.

100. Two outcomes are independent if and only if
    (a) they both lie along the least-squares line in a scatter plot
    (b) they are perfectly correlated
    (c) the occurrence of one outcome affects the probability that the other will occur
    (d) the occurrence of one outcome does not affect the probability that the other will occur
    (e) Wait! The premise is implausible. Two outcomes in an experiment can never be independent.

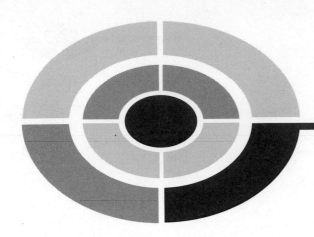

Answers to Quiz, Test, and Exam Questions

CHAPTER 1

1. b	2. b	3. b	4. d	5. c
6. d	7. b	8. a	9. c	10. a

CHAPTER 2

1. c	2. a	3. c	4. a	5. b
6. b	7. d	8. a	9. a	10. d

CHAPTER 3

1. a	2. c	3. d	4. a	5. c
6. b	7. d	8. c	9. b	10. b

Answers

CHAPTER 4

1. b	2. b	3. c	4. d	5. d
6. b	7. b	8. a	9. b	10. b

TEST: PART ONE

1. c	2. b	3. a	4. a	5. e
6. a	7. d	8. c	9. e	10. e
11. d	12. c	13. a	14. c	15. a
16. e	17. b	18. c	19. e	20. a
21. c	22. d	23. a	24. e	25. d
26. c	27. b	28. b	29. d	30. b
31. d	32. d	33. c	34. a	35. c
36. d	37. e	38. b	39. a	40. d
41. a	42. d	43. a	44. e	45. d
46. a	47. e	48. b	49. d	50. a
51. a	52. e	53. b	54. b	55. d
56. d	57. a	58. c	59. d	60. e

CHAPTER 5

1. b	2. a	3. a	4. a	5. a
6. d	7. c	8. b	9. d	10. a

CHAPTER 6

1. c	2. b	3. b	4. a	5. a
6. c	7. b	8. d	9. b	10. d

CHAPTER 7

1. a	2. b	3. c	4. c	5. a
6. a	7. b	8. d	9. c	10. a

CHAPTER 8

1. b	2. b	3. d	4. c	5. b
6. d	7. a	8. c	9. a	10. b

TEST: PART TWO

1. c	2. a	3. d	4. e	5. a
6. b	7. b	8. e	9. a	10. a
11. e	12. b	13. e	14. a	15. a
16. c	17. c	18. a	19. d	20. b
21. e	22. c	23. a	24. a	25. c
26. c	27. e	28. d	29. b	30. a
31. c	32. b	33. c	34. c	35. d
36. b	37. b	38. d	39. a	40. d
41. d	42. c	43. c	44. d	45. b
46. b	47. b	48. c	49. d	50. d
51. e	52. b	53. e	54. b	55. a
56. d	57. b	58. d	59. e	60. b

FINAL EXAM

1. b	2. a	3. c	4. a	5. b
6. a	7. a	8. b	9. c	10. a
11. c	12. e	13. a	14. e	15. d
16. c	17. c	18. b	19. a	20. a
21. e	22. e	23. c	24. c	25. b
26. a	27. b	28. d	29. b	30. c
31. c	32. d	33. c	34. d	35. b
36. e	37. b	38. b	39. b	40. c
41. b	42. a	43. d	44. a	45. d
46. b	47. d	48. d	49. e	50. c
51. e	52. d	53. b	54. b	55. e
56. e	57. c	58. e	59. a	60. d
61. d	62. b	63. c	64. c	65. c
66. e	67. a	68. a	69. d	70. a
71. c	72. d	73. c	74. c	75. c
76. b	77. e	78. b	79. c	80. c
81. b	82. b	83. c	84. d	85. a
86. e	87. e	88. a	89. b	90. e
91. c	92. d	93. e	94. a	95. b
96. d	97. c	98. a	99. e	100. d

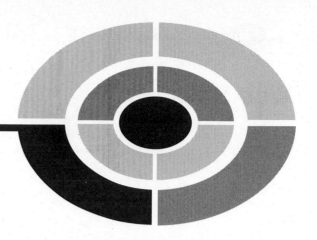

Suggested Additional References

Books

Downing, Douglas and Clark, Jeffrey, *Statistics the Easy Way – 3rd Edition*. Barron's Educational Series, Hauppauge, NY, 1997.

Graham, Alan, *Teach Yourself Statistics – 2nd Edition*. Contemporary Books, Chicago, IL, 1999.

Jaisingh, Lloyd, *Statistics for the Utterly Confused*. McGraw-Hill, New York, NY, 2000.

Moore, David S., *Statistics: Concepts and Controversies – 5th Edition*. W. H. Freeman & Co., New York, NY, 2001.

Stephens, Larry J., *Beginning Statistics*. Schaum's Outline Series, McGraw-Hill, New York, NY, 1998.

Web Sites

Encyclopedia Britannica Online, www.britannica.com
Eric Weisstein's World of Mathematics, www.mathworld.wolfram.com

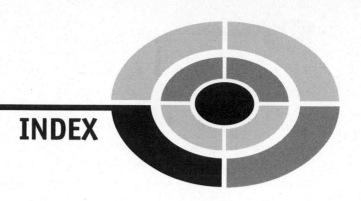

INDEX

ABOUT THE AUTHOR

Stan Gibilisco is one of McGraw-Hill's most prolific and popular authors. His clear, reader-friendly writing style makes his electronics books accessible to a wide audience, and his background in mathematics and research makes him an ideal editor for professional handbooks. He is the author of the *TAB Encyclopedia of Electronics for Technicians and Hobbyists, Teach Yourself Electricity and Electronics,* and *The Illustrated Dictionary of Electronics. Booklist* named his *McGraw-Hill Encyclopedia of Personal Computing* a "Best Reference" of 1996.